Geophysics and Environmental Physics

The series *Geophysics and Environmental Physics* (GEP), reports new developments in geophysics research and teaching with a high quality and the explicit aim to summarize and communicate current knowledge in an accessible way. Books published in this series are conceived as bridging material between graduate textbooks and the forefront of research.

The research fields covered by the series are:

- Atmospheric Physics, Environmental Physics, Biogeophysics and Physics of Climate;
- Earth Magnetism and Rock Physics, Seismology and Physics of Volcanoes;
- Geodynamics and Tectonics;
- Physical Oceanography and Hydrology;
- Surface Processes;
- Geodesy;
- Instrumentation and Data Systems;
- Nonlinear Processes in Geophysics;
- Planetary and Solar System Physics.

Proposal should be sent to a member of the Editorial Board, or directly to the Responsible Editor at Springer:

Dr Lisa Scalone - lisa.scalone@springernature.com

Jonathan D. Phillips

Abiotic Selection in Earth Surface Systems

Springer

Jonathan D. Phillips
Department of Geography
(Emeritus) University of Kentucky
Lexington, KY, USA

ISSN 2948-2186 ISSN 2948-2194 (electronic)
Geophysics and Environmental Physics

ISBN 978-3-031-85861-1 ISBN 978-3-031-85862-8 (eBook)
https://doi.org/10.1007/978-3-031-85862-8

This Springer imprint is published by the registered company Springer Nature Switzerland AG
The registered company address is: Gewerbestrasse 11, 6330 Cham, Switzerland

Preface

When Springer first approached me about writing a book for their Geophysics in the Environment series, I responded quickly with the idea for a book about selection in geophysical phenomena, independently of any biota on hand. I considered it a long shot as—despite, as will be revealed in this book, a considerable tradition of work on selection outside of Darwinian natural selection in biological evolution—many scientists reject selection as a guiding principle outside biology or applied metaphorically as analogous to Darwinian selection. Further, the idea that selection results in emergent phenomena better explains many Earth system phenomena that have long interpreted or explained in other terms has not necessarily been popular. But, given some initial encouragement from geoscientist/physicist Allen Hunt and Lisa Scalone of Springer, I developed a proposal—still considering it a long shot—because doing so would help me organize some of my thinking and arguments along these lines.

Lo and behold, they liked the idea, and here we are.

What I call the VSR (variation, selection, retention) logic may have occurred to any number of thinkers over the centuries but was most famously and influentially outlined by Charles Darwin. Variations occur, some of those are advantageous and are therefore selected for, and these are retained. As others have remarked, the principle seems simple and obvious once spelled out—a hallmark of great ideas. In biological evolution, the selection occurs due to better odds of surviving and reproducing and retention due to heredity (as we now know but Darwin did not at the time, via genes). Via natural selection, these ideas—quite appropriately—became baked into biology. Also baked in, though, was the idea that reproduction in the form of producing offspring and retention via heredity is the only "real" (as opposed to metaphorical) form of evolution by selection.

Positive selection, however, applies more broadly to forms, patterns, features, and processes that work better than alternatives. Retention can occur due to persistence of stronger, more stable, more resistant entities, and they often recur in multiple places and times (the *R* in VSR could also stand for replication, recurrence, or reinvention). Darwinian natural selection is one specific example of a much broader and more general phenomenon.

Much of my research has in fact dealt with the complex interactions among abiotic and biotic components of Earth surface systems and the coevolution of landforms, soils, hydrology, and ecosystems. Biological natural selection, ecosystem selection, and ecological filtering will not be ignored. This is because of the tightly interwoven connections between geophysical and biological processes, the role of natural and ecological processes in creating and facilitating (i.e., selecting) some geophysical processes and flux patterns, and as a pedagogical tool leveraging the more general familiarity with Darwinian selection. However, the latter will not be treated in depth, as other sources already do so, seemingly *ad infinitum.* My emphasis on geophysical and abiotic selection is to offset this imbalance, not to make a case for diminishing the role of biological and ecological selection.

Earth's biotic and abiotic components coevolve and are tightly interwoven, so drawing analogies between geophysical phenomena and natural selection is not unjustified. However, selection occurs in geophysical phenomena in a very real, not just metaphorical, sense. Further, abiotic selection can be derived and analyzed completely independently, as emergent phenomena. This book shows how selection in abiotic (as well as biotic) entities applies to geophysical processes and to the structure, function, and evolution of Earth surface systems. It demonstrates that phenomena traditionally interpreted in teleological or teleomatic terms or as goal functions (despite abiotic processes and features having no intentionality) are more realistically viewed and analyzed as emergent outcomes of selection. Abiotic selection in geophysical and geochemical processes is critical to understanding how Earth surface systems (ESS) function and evolve.

* * *

Beyond a general fascination with the incredible variety and complexity nature presents to us in Earth's topography, soils, climates, and ecosystems, my research efforts have usually been motivated by things I observed that I could not explain, which generally means that existing explanations struck me as inadequate.

As scientific fieldwork and measurements began to supplement a lifetime of poking around in the forests, swamps, marshes, and beaches of the southeastern United States, it struck me that there was often variability over short distances and small areas or volumes that could not be linked to any observable or even measurable variations in controlling or influencing factors. Thus began at interest in ways to quantify spatial variability. What I learned and devised helped, but often not in the way I had hoped, by enabling direct connections between process and response and between environmental controls and observed spatial patterns. Rather, in many cases it led to the realization that *something* was causing properties of Earth surface features to diverge spatially in ways that are disproportionately large relative to whatever caused or controlled their development or to last much longer than any evidence of whatever triggered their developmental divergence. Seeking a way to conceptualize this fact led me to dynamical instability, deterministic chaos, and other theories and methods associated with complex systems.

As most others working on complex systems in the geosciences at the time, I relied on methods and concepts imported from mathematics, systems theory, and physics. These were adequate to convince me that the (heretofore) unexplained variations could be explained by dynamical instability and chaotic exaggerations of the effects of minor disturbances and initial variations. However, many of my Earth and ecosystem science colleagues, if not wholly skeptical, were unsatisfied, and their response highlighted some of my own doubts. Much of my (and everybody else's) evidence came from mathematical models—reasonable and plausible ones, I like to think, but as Ian Stewart and Martin Golubitsky put it in their book *Fearful Symmetry*: "Mathematically, you can balance a chain end to end, provided each link sits precisely above the one below it .... This is impossible in the real world. So mathematical *existence* ... isn't enough." Mathematical existence isn't enough for Stewart and Golubitsky, mathematicians both, and it damn sure isn't for your average field scientist. "Enough" requires us to move beyond identifying the presence of deterministic complexity to address the processes and mechanisms producing it and the observable effects in topography, stratigraphy, matter and energy fluxes, soil morphology, weather and climate patterns, and so on. It also involves problematizing ESS development from *within* geosciences and ecosystem sciences, rather than posing problems of, e.g., climate patterns, hydrological flow systems, and landforms as mathematical or theoretical physics questions.

Our collective work along these lines eventually showed that dynamical instability and chaos, for example, are real and important in ESS, though certainly not evident everywhere, always, and at every scale, even within a single system. The mechanisms behind the deterministic complexity—e.g., chaos in fluid turbulence, unstable wetting fronts, divergent evolution of topography—have been identified for many phenomena. This leads inexorably to the conclusion that where minor variations in initial conditions (impossible to observe in many Earth science problems) or small disturbances can have disproportionately large and long-lasting impacts, ESS are *idiosyncratic*. Place and history matter, and we cannot explain ESS without accounting for the local environmental and historical context.

This suggests that what many of us were taught was the holy grail of science—to eventually explain everything based on generally applicable laws and principles that are independent of place and time—is impossible. When presenting some of these ideas in the context of soil and regolith development, more colleagues than one objected with something along the lines of: "Yes, but if you get me enough detailed information on each profile, I can explain it entirely on the basis of general laws." I agree—me too. But the point is, *that's what you would have to do!*

I never doubted that, given sufficient local detail and suitable analytical and measurement techniques, any given ESS could be explained mostly, if not entirely, based on generally applicable laws and principles. However, I also do not doubt that is what must be done to achieve explanation based only on general laws, and that those explanations themselves would not be generalizable.

On the other hand ....

Recurring forms, patterns, and arrangements certainly exist in nature. The common features of soil profiles, stream channel networks, beach morphologies, and so on suggest some underlying law or generalization that guides their formation. The whole subfields of synoptic meteorology and climatology are based on the presence of recurring forms and patterns. But no physical law dictates the formation of beach cusps or makes tropical cyclone formation inevitable. And while the topological properties of channel networks and vertical differentiation of soils are widespread, they are not universal. There certainly exist invariant physical laws governing sediment transport capacity of streams, but none that enforce a balance between sediment supply and transport capacity. As I once heard geomorphologist Dick Marston say, "rivers don't *try* to do anything."

Neither do rocks, hillslopes, soils, or carbon dioxide molecules.

This is what led me to *selection,* and I was hardly the first to get there. Many of the ESS features and phenomena on our planet recur or persist because they are more stable or resistant than others. Or because they work better (more efficiently) for energy and matter fluxes and deplete gradients faster. Or because they are advantageous for the functioning of ecosystems. There are obvious analogies with Darwinian natural selection, but as those geoscientists who got here earlier have made clear, one need not reason or argue by analogy. Selection happens independently in abiotic processes, and Darwinian natural selection is but one member of a large family of selection processes.

In this book, I have sought to highlight the ubiquity of abiotic selection in ESS. I have also tried to give numerous examples from disparate geophysical phenomena to illustrate *why* certain forms and patterns are advantageous, and *how* they arise and are selected for. Finally, I have made a case for emergent properties arising from selection as a broad explanation for many observations in ESS.

* * *

Hopefully, the intellectual debts I owe for the material in this book will be clear enough from the discussions and citations herein. I will single out Gerald Nanson and Allen Hunt, however, who in some cases have quite different ideas and perspectives on the topics herein. Through various interactions, not infrequently involving reviews of each other's manuscripts, I learned that in some cases our views were not quite as different as I first thought, and that even where they are different, they are not necessarily incompatible. Working through these, from well before I started this writing project, has been critical for clarifying my thoughts on abiotic selection.

I appreciate Springer's support for this project, and Lynn Roche Phillips' understanding and indulgence for my immersion in this project while supposedly being retired.

New Bern, NC, USA
January, 2025

Jonathan D. Phillips

Competing Interests The author has no competing interests to declare that are relevant to the content of this manuscript.

Contents

Introduction: Abiotic Selection in Earth Surface Systems

1

Abstract

The role of selection in Earth surface systems (ESS) is widely recognized in the form of Darwinian natural selection. Biotic-abiotic interactions are evident in ecological filtering and ecosystem engineering. But selection principles also apply to geophysical, abiotic components of ESS. Natural selection an example of a wider principle summarized by variation-selection-retention (or recurrence)—the VSR logic. Application of this principle to ESS involves biotic, abiotic, and hybrid phenomena. Abiotic selection has been far less recognized and studied and is emphasized here. Emergent phenomena arising from selection are proposed as explanations for the seemingly contradictory presence of unlimited variability and recurrence of certain forms and patterns. This chapter reviews ideas about selection in geoscience contexts other than Darwinian natural selection, and provides three examples of abiotic and geophysical selection in atmospheric heat flux, form-process relationships in alluvial stream channels, and evolution of soil hydrologic systems.

1.1 What This Book Is About

This book shows how selection in abiotic (as well as biotic) entities applies to geophysical processes and to the structure, function, and evolution of Earth surface systems (ESS). It demonstrates that phenomena traditionally interpreted in teleological or teleomatic terms, or as goal functions are more realistically viewed and analyzed as emergent outcomes of selection. Selection and adaptation occur everywhere and always, whether biotically or physically mediated. All forms of selection are probabilistic. This means any rules developed from theoretical treatment of selection principles can be expected to explain only the ubiquity of certain

J. D. Phillips, *Abiotic Selection in Earth Surface Systems*, Geophysics and Environmental Physics, https://doi.org/10.1007/978-3-031-85862-8_1

generalizable patterns in nature, not their precise description, nor what variation will occur in any particular locale.

Earth Surface System (ESS) is an umbrella term for landscapes and environmental systems operating at or near Earth's planetary surface. It includes geological and geomorphic, pedological, hydrological, and climate systems as well as ecosystems.

Think about it: all ESS are governed by some general, universal laws and principles that operate everywhere and always—either in a literal sense or within certain domains. Despite these common rules there exists an unlimited variety of landforms, landscapes, geological formations, soils, hydrological flow systems, weather systems, ecosystems, and so on. Now think about it another way: Earth presents a wide variety of macro- and microclimates, geological substrates, biological influences, topographic and bathymetric settings, geological histories, and so on. Yet—though no two rivers or glaciers or ocean circulation zones or rock formations are exactly the same—there exist some remarkable similarities and recurring patterns across the planet. How to explain the coexistence of these apparently contradictory phenomena?

To some extent such regularities as do occur are attributable to the general laws. But even where the same set of laws operate, they often do so at different rates and intensities, with different degrees of relative importance, and in different environments that constitute varying boundary conditions, energy and matter resources, and system feedbacks. In this book I show that the constrained variation of ESS—the mixture of idiosyncrasy and similarity—is due to the operation of *selection* processes.

Natural selection in biological evolution has produced an increasingly diverse and complex biosphere. It is also widely appreciated that no matter what mutations occur or what evolutionary innovations arise, there are rules and constraints based on laws of chemistry, physics, and biology. We also know that biological evolution is probabilistic and not deterministic—the fittest do not *always* survive and reproduce, and the less fit sometimes do.

A similar story can be told about evolution at the level of ESS, but not only due to the coevolution of biota and other aspects of Earth systems or the effects of organisms on the abiotic environment. Selection also applies to geophysical and other abiotic processes and features.

Non-biological and *abiotic,* as used in this book, do not necessarily refer to purely abiotic phenomena with no biological influences. Such phenomena are quite rare, as the biosphere has direct or indirect influences on almost everything, and even the harshest environments in the Earth-atmosphere system have something living in them, even if only microbial extremophiles. But there does exist a continuum of biological influence, from essentially biotic constructs such as coral reefs and peat bogs, to lava flows, hyperarid sand dunes, and glacial grooves where biological influences are indirect and trivial. Many geophysical processes, such as fluid flows of water and air, can generally be studied independently of any biological influences, or have aspects that are independent of organisms. Thus, for instance, when we speak of geophysical fluid dynamics in the atmosphere, we are not claiming that biota are totally disconnected—composition of the fluid in question, the air, is largely controlled by the biosphere. Rather, speaking of the dynamics as abiotic reflects the fact that they are not directly influenced by organisms and can be understood independently of biology.

Selection is perceived by most scientists to refer to Darwinian natural selection in biological evolution, perhaps with extensions to ecology, or to phenomena closely analogous to Darwinian natural selection (hereafter simply natural selection). But natural selection is only one example of a much broader range of selection dynamics in nature, including abiotic phenomena. Abiotic selection in geophysical and biogeochemical processes is critical to understanding how ESS function and evolve.

Earth's biotic and abiotic components coevolve and are tightly interwoven, so drawing analogies between geophysical phenomena and natural selection is not unjustified. However, selection occurs in geophysical phenomena in a very real, not just metaphorical, sense. Further, abiotic selection can be derived and analyzed completely independently, as emergent phenomena. The fundamental logic is: (1) *Variations* occur; (2) some variations are more efficient, stable, durable, or otherwise favorable and are therefore *selected* (3) the latter are more likely than others to be *retained,* via preferential preservation, enhancement, or replication. This is sufficient for selection to occur. For convenience we can think of this as the *variation-selection-retention* or VSR logic. Very similar logic was proposed by Wong et al. (2023) as a law of nature on the same level as invariant physical laws, though the basic ideas of VSR logic go back much farther (Sect. 2.1.1).

This book is focused on geophysical selection, but natural selection, ecosystem selection, and ecological filtering will not be ignored. This is because of the tightly interwoven connections between geophysical and biological processes, the role of biological and ecological processes in creating and facilitating (i.e., selecting) some geophysical processes and flux patterns, and as a pedagogical tool leveraging the more general familiarity with Darwinian selection. However, the latter will not be treated in depth, as other sources already do so, seemingly *ad infinitum.*

Geophysical selection takes four general forms: Gradient, resistance, network, and thermodynamic selection. Each of these will be explored in detail in later chapters.

1.2 Evolution and Selection

Some evolutionary biologists object to the use of terms and concepts of evolution and selection in any context other than biological evolution, insisting that where there are no genes involved there can be no selection or evolution. Other scholars insist that to evoke evolution and selection in any context other than the one Darwin and Wallace pioneered requires a close analog between the context in question and biological evolution. However, Darwinian evolution by natural selection is a specific example of a broader phenomenon.

The word evolution may simply refer to development or change over time, not necessarily implying anything else. It is sometimes used to draw a contrast with concepts or words such as development, succession, or progression. The latter imply deterministic progression along a single pathway toward a single end point, whereas evolution signifies the possibility of multiple pathways and outcomes, and a potential role for chance and path-dependence. In this book evolution refers to systematic but non-deterministic development over time that is often historically contingent and path dependent.

In dealing with abiotic and geophysical selection there is no need for arguments by analogy. Selection principles are directly applicable to many physical, chemical, and system-level (as opposed to organism level) phenomena in ESS. E.N. Lorenz (1960) may have been the first to explicitly link selection to geophysical processes. He argued that patterns of atmospheric heat flux self-organize to maximize the rate of potential energy production via selection. G.W. Paltridge (1975) found selection operating on ocean-atmosphere heat exchange to minimize entropy exchanges with the external environment.

Adrian Scheidegger proposed five fundamental principles of landscape evolution, including the *selection principle* (Scheidegger, 1987, 1991). As landscapes evolve by denudational processes, the most geostatically stable are selected by weathering and erosion processes by removal and transformation of the less stable. Thus, the surviving forms and configurations are those that are most stable under their own weight. This is a form of the *resistance selection* covered in Chap. 5. Scheidegger pointed to Gerber's (1969) study of alpine glacial morphology as the origin of the concept.

Luna Leopold (1994) described river morphology and its changes as an analog to Darwinian natural selection, but Gerald Nanson and He Huang went farther in invoking selection in fluvial geomorphology and hydrology independent of any Darwinian analogs. In a series of papers beginning in 2000, Huang and Nanson connected the least action principle to selection in geomorphological and hydrological systems. We'll examine their ideas in more detail below.

By the 2000s, as we will see, selection was a relatively common theme in studies of abiotic aspects of ESS—occasionally via analogies with Darwinian natural selection, but mostly independent of biological analogies, based implicitly or explicitly on VSR logic. First, however, let's review selection ideas that increasingly generalize from genes and individual organisms to ESS and abiotic phenomena.

Note that in biology, abiotic selection often notes selective effects exerted by abiotic factors on natural selection of organisms. These selective effects of abiotic factors are not irrelevant to our task here, but the focus is on selection *of* abiotic factors, independently of biological selection—that is, selection acting on geophysical entities rather than geophysical entities exerting selection pressures on biota.

The biotic/abiotic dichotomy is, if not artificial, at least highly exaggerated. The atmosphere, lithosphere, and hydrosphere do not exist or function independently of the biosphere (or of each other). But selection in the abiotic sector has been underappreciated, and even denied, so I seek here to redress the imbalance.

1.3 From Darwinian Natural Selection to Geophysical Selection

1.3.1 Natural Selection

Global production of maize (corn in the U.S.) is about 1.2 billion tonnes per year, with the global maize market estimated to be nearly US$144 billion in 2024. This all started roughly 9000 years ago when humans in what is now southern Mexico began to domesticate a grass called teosinte (*Zea mays parviglumis* and *Z. Mexicana*). By selectively breeding plants with desirable traits and discarding those with undesirable ones, over the years the tough-seeded teosinte was transformed into its ancestor, *Zea mays mays*—maize or domesticated corn. Only an expert botanist could detect the links between our maize and its ancestor. This was the result of deliberate selection, of the type Darwin termed *artificial* selection, to contrast it with his idea of *natural* selection.

Darwin realized that nature could also select individuals, albeit in a more drawn-out and haphazard way. Individuals better suited to utilize the available resources, weather the local stresses, and succeed in reproducing are more likely to survive, thrive, and produce progeny. If the traits that make them better suited are heritable, they can be passed along to the next generation—and evolution happens. The essential idea of biological evolution by natural selection is well established and not seriously challenged, on scientific or logical grounds, anyway. The mechanistic details of evolutionary genetics and changes in complexity are still unsettled, as is the extent to which evolutionary change occurs gradually and incrementally, as opposed to episodic bursts of change interrupting long periods of stasis. But evolution by natural selection indeed happens, one way or another.

Natural selection is not deterministic, but probabilistic. The fittest, healthiest, smartest, strongest, most genetically blessed individuals are not always, invariably the ones that survive the longest and reproduce the most. Weaker, dumber, more poorly suited, and less genetically advantaged individuals do sometimes survive and reproduce. The fitter or better adapted pass their genes along on average, but not in every case. Genetic mutations (some advantageous, some not) and genetic drift also bring luck into play.

1.3.2 Ecological Filtering

Ecological filtering is a form of selection strongly influenced, and often dominated, by the abiotic environment. Organisms have finite ranges and tolerance limits for the stresses they are subjected to and for the resources they require or can utilize. The environment selects by filtering out non-adapted or unfit individuals by preventing or impeding their establishment, health, and reproduction. The abiotic environment also allows or facilitates establishment of better-suited organisms. For example, estuarine wetlands filter out plants that are intolerant of soil saturation, salinity, and flooding, and encourages those well adapted to wetness, salt water, and flooding. Ecological filtering is not restricted to abiotic factors. A dense forest canopy, for example, allows shade-tolerant and filters out shade-intolerant plants.

Reciprocal, coevolutionary dynamics also exist. Organisms modify their abiotic habitats through ecosystem engineering, thereby affecting ecological selection pressures (positively or negatively) for the engineer species and others. Natural selection may also be directly affected by ecological filtering. Adaptive diversification driven by "ecological opportunity" relates to the environmental resources and stresses. The spatial and temporal structure of environmental factors, along with the biological properties of a lineage that determine its capacity to diversity, are the primary controls of evolutionary responses to ecological opportunities such as new environmental conditions (Wellborn & Langerhans, 2015).

1.3.3 Biological Evolution as a Physico-Chemical Process

The U.S. National Aeronautics and Space Administration's (NASA) astrobiology program defines life (on Earth) as "a self-sustaining chemical system capable of Darwinian evolution" (https://astrobiology.nasa.gov/research/life-detection/about/). Vitas and Dobovisek (2019) suggested an extended version including thermodynamics. They hypothesized that as life emerged, a first step was evolution of autocatalytic systems that only later acquired the capacity of genetic heredity. Their definition is: Life is a far from equilibrium self-maintaining chemical system capable of processing, transforming and accumulating information acquired from the environment. This extends the definition to allow for non-Darwinian forms of selection and does not distinguish between the metabolic and ecological levels. Their work builds on earlier frameworks going back at least to Kauffman (1971; see also Kauffman, 1996) positing that autocatalysis and other forms of self-organization made the evolution of life inevitable.

Life is supported by, and can be reduced to, a multitude of chemical reactions, Sharma and Annila (2007) asserted. They formulated entropy for an open thermodynamic system undergoing chemical reactions, asserting that the fitness benchmark is movement toward more probable states characterized by greater entropy production by various mechanisms. The quest for more probable states, they argued, results in organization of matter in functional hierarchies (Sharma & Annila, 2007). The question of how or why chemicals, as opposed to organisms, should "quest" for anything is not addressed. However, it is implied that evolution by natural selection is subsumed by the second law of thermodynamics, a point made explicit by Annila and Salthe (2010), who hold that selection favors the fastest reduction of energy differentials. Kaila and Annila (2008) proposed that natural selection and the principle of least action both deal with natural motions and reconcile to one law. The second law written as a differential equation of motion describes evolution along the steepest energy gradients. In integral form, motion is depicted to occur along the shortest paths in energy. In general, evolution is a non- Euclidian energy density landscape in flattening motion. Due to random variations, matter, energy, and information flows explore different evolutionary paths. Those that transfer energy from higher to lower states (equivalent to a faster entropy increase) are selected. As discussed below, others have also explored links between the second law/least action principle and selection.

Stability is a logical connection between the physical and biological worlds, Pascal and Pross (2016) proposed. Time is the most general aspect of stability, manifested as persistence, leading them to the persistence principle: nature seeks persistent forms. They later (Pross & Pascal, 2017) developed the idea in a more specific form based on the thermodynamics of chemical kinetics. Dynamic kinetic stability (Pross & Khodorkovsky, 2004) is based on time or persistence rather than free energy and, Pross and Pascal asserted, provides a basis for understanding evolution. Kinetics is important in biological evolution, and in non-living systems all selection is thermodynamic. Thus, the same principles govern physical and biological systems (Pascal & Pross, 2016; Pross & Pascal, 2017).

Egbert et al. (2023) proposed that antecedents of life were already actively regulating the environment to support their own persistence, a phenomenon they called *viability-based behavior*. They presented four non-biological systems that demonstrate self-preserving behaviors: reaction-diffusion 'spots,' motile oil-droplets, ramified charge-transportation networks, and Bénard convection cells. These systems move or reconfigure themselves in a way that changes their environment (e.g., by moving to a different location) or modifies the way that it interacts with its environment, in a way that increases the likelihood of its persistence.

1.3.4 Ecosystem Evolution

Do evolution and selection occur at the ecosystem level?

Arguments for and demonstrations of ecosystem evolution can be linked to the variation-selection-retention logic. Variations in modes or patterns of resource use (e.g., energy fluxes, water use, or biogeochemical transformations) exist. Some of these variants improve the odds of survival and reproduction, and/or functional capabilities, for individuals, communities, or flux networks. Those advantage-conferring structures, relationships, or modes are selected for (preferred). Nothing else is necessary to account for directional evolutionary trends in ecosystems such as increasing rates of water, energy, or biogeochemical cycling, which can be linked to greater efficiency and entropy production (Smith, 1986; Eagleson, 2002; Lapenis, 2002; Lekevicius, 2002, 2006; Phillips, 2008, 2021).

Ecosystem evolution is generally accepted by paleoecologists, both with respect to generic types of ecosystems (e.g., semi-arid grasslands, tundra, tropical rainforests) and to specific ecological systems. The paleoecological record of ecosystem change shows that changes in biological communities and their associated abiotic environments occur in response to external forcings or changes in boundary conditions, and to internal dynamics. Paleoenvironmental archives show that disturbances (e.g., volcanism, fire, catastrophic storms) and environmental change (e.g., sea-level, climate) can trigger and influence the trajectories of ecosystem change. For example, a review of grassland ecosystem evolution (Strömberg, 2011) takes into account evolution of the grasses, coevolution of plants, herbivores, and soils, and triggers of change such as fire, climate, atmospheric chemistry, and ecological interactions with woody plants. This is representative of paleoecological perspectives on ecosystem evolution in acknowledging the interplay of abiotic change, biological evolution, and biotic-abiotic interactions.

In their study of plant communities on the Colorado Plateau, Duniway et al. (2022) identified, even before interactions among components are considered, a chain of selection from depositional environments that determine rock characteristics, to weathering, erosion, sedimentation and other aspects of geomorphological development, to parent material and topographic effects on soil formation, to edaphic effects of soils on vegetation.

In the context of biogeomorphology and ecosystem engineering, Corenblit et al. (2010) distinguished between abiotic components of landforms strictly controlled by physical processes; residual components derived from biota due to niche-constructing activity or ecosystem engineering, but with but no apparent benefit to the engineer species; and functional components that do benefit the engineer or other species.

1.3.5 Abiotic and Geophysical Selection

"Before plants and animals invaded the land," Murray et al. (2008) wrote, scientists "would have beheld a landscape adorned with loose sediment: naked clasts making their way down highly incised hillslopes, sand and silt blowing free or piling up in ubiquitous dunes, most rivers braiding with abandon." Geoscientists often pine for such a field area (or seek them out in arid, arctic, or alpine settings) "where physics rules without the influence of biology." Alas—or fortunately, depending on one's perception—Earth systems generally involve some biological and ecological influences that oppose the complete rule of deterministic physics.

The purported geophysicist's dreamland might be fairly tame in some senses. Vernadsky (1926) saw the biosphere as a planetary membrane for capturing and converting solar energy, much of which would otherwise simply be returned to space. Phillips (2009) showed that even conservative estimates of biological energy subsidies to landscape evolution greatly exceed energy derived from geophysical processes. Dyke et al. (2011) estimated that biota generate 264 terra watts (TW) of power annually, much greater than geophysical processes in the solid Earth (mantle convection was estimated at 12 TW, for instance). Only a small fraction of the biologically harvested solar energy needs to be used for geomorphic impacts for the impacts on surface processes and landforms to affect interior geological processes (e.g., via isostasy).

However, selection applies to dominantly geophysical processes as well as the confounding organisms. In mechanics the least action principle (LAP) states that the motion between any two points in a conservative dynamical system is such that the action has a minimum value with respect to all paths between the points that correspond to the same energy. This is often summarized as "nature finds or chooses the most efficient path." The LAP can be stated for the case of a moving object as the average kinetic energy minus the average potential energy is the minimum possible for the path from one point to another. Least action is manifested in ESS by the accomplishment of work—for example heat flux, ecosystem productivity, sediment transport—using as little of the available energy as possible (this implies maximizing entropy production, a link we will explore later). For a rocket ship, baseball, or cannon ball a path of motion according to the LAP is deterministic and straightforwardly calculated. A simplistic link with selection is the idea that nature selects or chooses the most efficient path. However, selection is also associated with the LAP in more subtle and complex ways, as we will explore in Chap. 2.

Generalized selection principles for flows in physical systems are formalized in the constructal law: For a flow system to persist in time (to survive) it must evolve in such a way that it provides easier and easier access to the currents that flow through it (Bejan, 2007; Bejan & Lorente, 2013; Lorente & Bejan, 2019). This has been applied to phenomena including wedge-shaped turbulent shear layers, jets and plumes, the frequency of vortex shedding, Bénard convection in fluids and fluid-saturated porous media, and coalescence of solid parcels suspended in a flow. Global ocean and atmospheric circulation are also purported to reflect the constructal law. Bejan (2017) argued for a direct, explicit link between evolution/selection and thermodynamics, such that thermodynamic systems (essentially, all systems that move or change) evolve—a theme explored further in Chaps. 2 and 3. If selection leads to persistence of the preferred flow paths, constructal-type behavior is obtained via traditional path-of-least-resistance reasoning.

Additionally, Egbert et al. (2023) proposed ice-albedo feedbacks as an example of the viability-based behavior in non-biological systems that parallels Darwinian evolution. Increased sea and land ice cover increases albedo, and more solar radiation is reflected. This cooling effect helps maintain the conditions necessary for ice cover persistence and growth. This implies that any self-reinforcing positive feedback constitutes viability-based behavior that constitutes evolutionary advantages for the positively self-reinforced entity; an idea we will explore further in Chap. 2.

1.4 A Selection of Selection in Earth Surface Systems

Here we provide some brief examples of abiotic selection in Earth surface systems.

1.4.1 Atmospheric Heat Flux

G.W. Paltridge (1975) presented a model for climate thermodynamic zonation based on maximum entropy production (MEP). Annual climate of the Earth-atmosphere system should be such that thermodynamic dissipation, associated with the horizontal fluxes in the ocean and atmosphere, is at a maximum, Paltridge reasoned. Selection comes into play as, from a large variety of possible stable states, the one that maximizes entropy production is selected.

In the model the Earth-atmosphere system is represented as a set of cells, each of which is characterized by mean parameters that determine the atmosphere-ocean state in terms of total sensible and latent heat, horizontal energy fluxes, surface temperature, and cloud cover (Fig. 1.1). The system is assumed to be stationary, with two energy balance equations applying to each cell. A complete energy balance for the cell is described by one equation, which depends on the flux of solar energy into the cell minus energy leaving it to the upper atmosphere and space; and on meridional fluxes between adjacent cells in the ocean and atmosphere. The second equation describes the ocean energy balance, accounting for insolation and

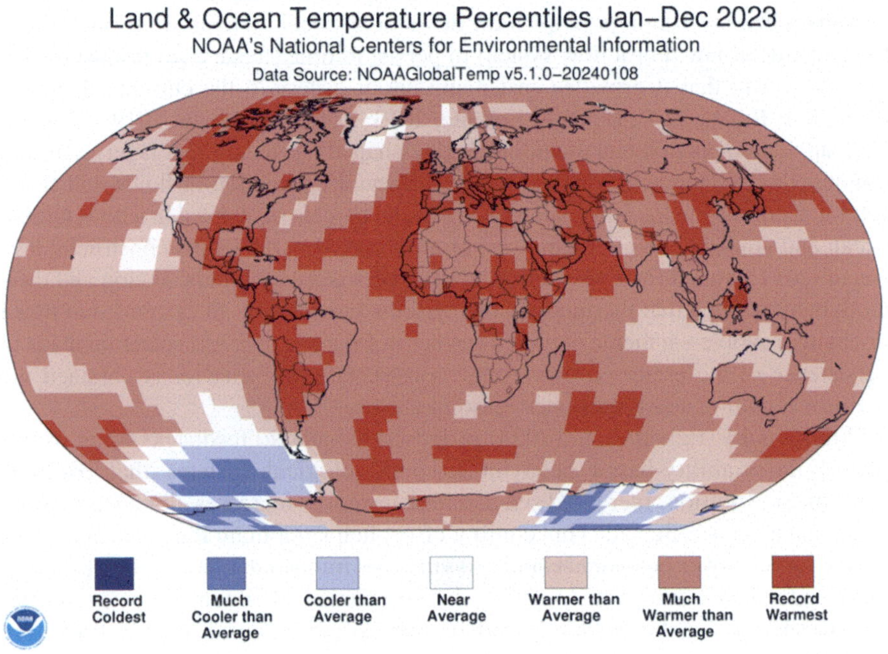

Fig. 1.1 U.S. National Oceanic and Atmospheric Administration map showing temperature anomalies in 2023 by grid cells. The prevalence of pink to red colors reflects that 2023 was the hottest year on record. Paltridge's framework asserts that the energy balance in each cell selects, from all potential states, the one that maximizes the rate of entropy production. (Reprinted from https://www.ncei.noaa.gov/access/monitoring/ghcn-gridded-products/#global-maps-select. © 2024, NOAA)

absorption by liquid water and water vapor, outflux from the sea surface, and the meridional fluxes mentioned above. Paltridge (1975, 1978, 1981) gave explicit expressions in the original papers. These equations are not sufficient for determination of all the cell characteristics, so additional assumptions are necessary, the main one being a requirement that total horizontal heat fluxes in all cells maximize the entropy production σ—that is, that the MEP state is *selected* from among the available states:

$$\sigma = -\sum_i \left(F_{S,i}\right) - \left(F_{L,i}\right) / T_i^* \tag{1.1}$$

$F_{S,i}$, $F_{L,i}$ represent energy fluxes arriving at the *i*th cell from the sun, and leaving the cell. T_i^* is the mean temperature in the *i*th cell, and the summation is over all cells. Due to the stationarity assumption, the total entropy produced in the system and the entropy flux across the boundaries are equal.

Notation note: throughout this book the symbols and variables in equations generally match those in the original sources, and are defined when they are first used in a section. Therefore, there is consistency within sections, but between sections the same symbol may have different meanings.

The relatively simple model predicted average annual distributions of temperature, cloud cover and heat fluxes on Earth that agree well with observations and became widely accepted in studies of planetary climates (Ozawa et al., 2003; Martyushev & Seleznev, 2006). The maximum entropy framework is not without critics and will be addressed in more detail in Chap. 7.

1.4.2 Alluvial Channels

Alluvial stream channels are a canonical example of mutual adjustments in ESS—channels are formed and modified by the flows they convey, while flow dynamics are affected by characteristics of the channel. Multiple "extremal" hypotheses have been advanced to explain alluvial channel development and changes, based on maximizing or minimizing some parameter, including maximum energy loss, sediment transport capacity, flow resistance, friction factor, and rate of energy dissipation, or minimizing variance, energy dissipation rate, stream power, and Froude number. These we will revisit in Chap. 7, but for now we focus on the body of work most directly and explicitly related to selection—the least action principle-based work of He Huang and Gerald Nanson.

The least action principle in rivers is manifested by maximum flow efficiency (MFE) and is based on variational mechanics of energy and work (as opposed to Newtonian force and momentum). Flow in most rivers consists of imposed water and sediment loads transported down valley gradients that are largely inherited. The rivers self-regulate their energy expenditure, Nanson and Huang argue, to match the work necessary to remain stable. Fluvial systems with excess energy to dissipate develop a variety of channel patterns and morphologies to become or remain stable. Selection comes into play as the stable configurations that maximize flow efficiency in the constantly-changing alluvial channel are the ones that preferentially recur and persist—survival of the most stable. With MFE, channels have just the energy needed to transport their water and sediment loads. Excess energy must be expended (dissipated) or the channel will become unstable. Channels with insufficient energy will deposit sediment and also become unstable; stability defined in this case as minimal change.

Flow continuity—conservation of matter and energy—governs fluid motions in channels. Of the many possibilities that may satisfy these laws, according to the Bélanger-Böss theorem of critical flow (Bélanger, 1849; Böss, 1919) there exists a unique minimum-energy solution. In Fig. 1.2, the minimum specific energy (potential + kinetic energy) occurs at the critical flow boundary between sub- and supercritical flow. This minimum specific energy condition in open channel flow indicates a Froude number (F_r) of one:

$$F_r = V / \sqrt{gD} = 1 \quad (1.2)$$

The Froude number represents the balance between velocity (V, the inertial force of the fluid) and the gravitational force of the fluid (D = depth of flow; g is the gravity constant). The forces are balanced when $F_r = 1$. Flow disturbances, such as those produced by objects or irregularities in the bed, vegetation, and debris, produce forces in the form of gravity waves that radiate from the location of disturbance. This can be visualized for the case of the circular ripples resulting from a stone cast

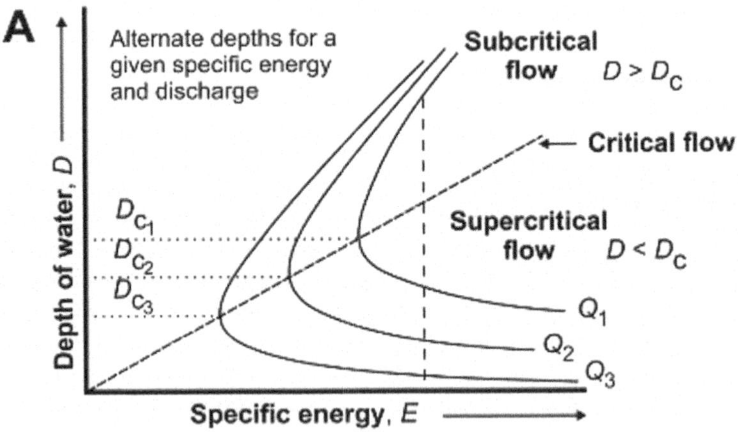

Fig. 1.2 Generalized specific energy diagram showing minimum specific energy (potential plus kinetic energy) at three critical depths D_c corresponding to three discharges, $Q_1 < Q_2 < Q_3$. The vertical dashed line shows that for a given amount of energy there can be two different depths, at sub- or supercritical flow. The inclined dashed line indicates only one minimum specific energy state for each discharge, where the Froude number is unity. (Reprinted with permission from Nanson and Huang (2017). © 2016 John Wiley & Sons, Ltd. All rights reserved)

into a pool of still water. In fluvial channels, downstream flow (inertial forces) slows the passage of any upstream gravity waves. When V equals the celerity of any gravity waves generated, $F_r = 1$ and energy cannot be expended upstream. This is *critical flow,* and any additional increase in V will cause the flow to change to supercritical, where inertial forces driven by kinetic energy dominate.

Though widely applied in open channel hydraulics, bedload sediment transport and flow resistance at the boundary complicate use of the Froude relationship to identify optimum flow conditions. The relationship is strictly applicable only to frictionless channels without bedload; conditions not met in river channels. Nevertheless, the Bélanger-Böss theorem implies the existence of an inherent law of motion in open-channel flow with friction and bedload transport.

A straight reach with a trapezoidal cross-section is the simplest case of a self-adjusting alluvial channel. Three basic flow equations (continuity, flow resistance, sediment transport) control the flow, with four variables (V, D, width, slope), precluding mathematical closure. The extremal hypotheses have been proposed to solve this problem, by assuming that, at equilibrium, flow maximizes or minimize some property. Huang and Nanson attacked the problem by collapsing width and depth into a single channel form factor—the width/depth ratio, ζ. They identified the energy gradient of water or sediment discharge as a function of ζ, making no *a priori* assumptions. They showed that for a given bedload Q_s and energy grade slope S there are typically two possible dynamic equilibrium forms of channel geometry that can transport a given amount of water and sediment with no channel erosion or accretion (Fig. 1.3).

At either point (A, A' in Fig. 1.3) the channel would be in mass-balance equilibrium. But only at the turning or stationary point of the curve does there occur a unique slope and *W/D* ratio where total energy expenditure is minimum—the stationary equilibrium where the imposed sediment load is transported with the least amount of energy. This is the MFE state, defined as the maximum sediment transporting capacity per unit of available stream power, with cross-section stream power Ω given by

$$\Omega = \gamma Q S \tag{1.3}$$

where γ is the specific gravity of the fluid and Q is the water discharge. Analogous to F_r in the Froude analyses, Huang and Nanson developed a dimensionless parameter H representing the ratio of relative increment of shear stress above the critical shear stress for bedload transport to the increment of W/D above the optimum for conveying pure water. $H = 0.3$ when alluvial channel flow is at the stationary equilibrium state:

$$H = \frac{\Delta\tau_c / \tau_c}{\Delta\varsigma} = \frac{\tau_o / t_c - 1}{\zeta - \zeta_{c\zeta}} = 0.3 \tag{1.4}$$

where τ_o and τ_c are, respectively, the mean boundary shear stress and the critical shear stress for bedload transport, and ζ and ζ_c are the optimum W/D ratio and the optimum critical W/D ratio when mean boundary shear stress is equal to the critical value, or when Q_s is zero. $F_r = 1$ when inertial and gravitational forces are balanced;

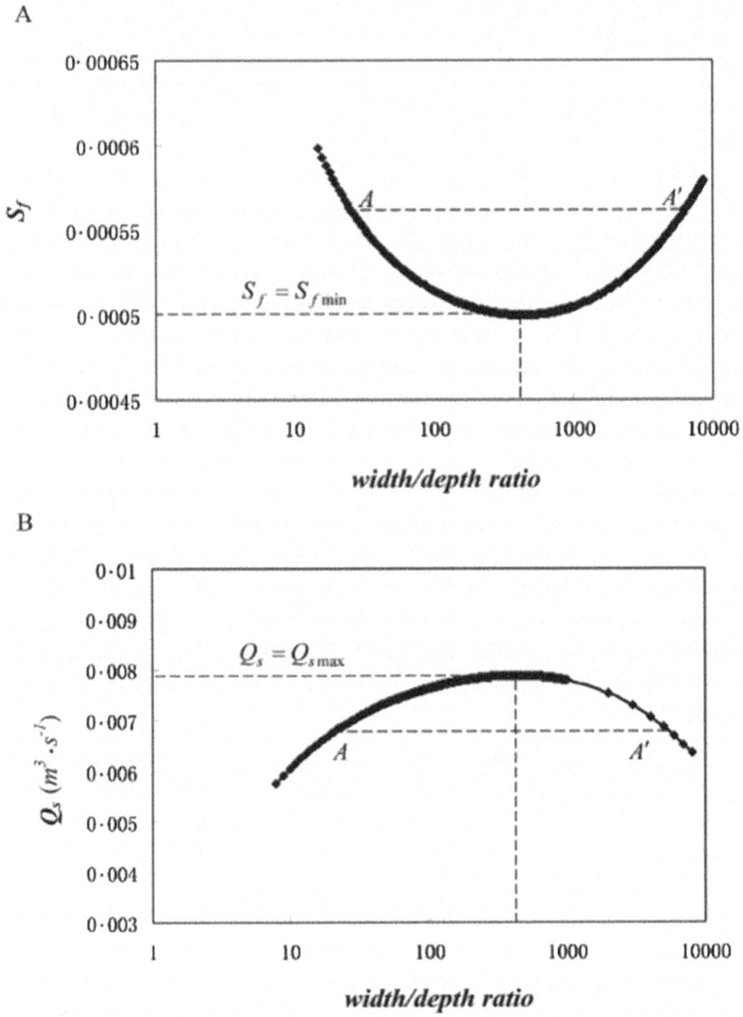

Fig. 1.3 Worked example based on the theoretical relationship developed by Huang and Nanson (2000, 2007) and Huang et al. (2004) of variations in sediment transport Q_s with changes in width/depth ratio ($Q = 100\ m^3\ s^{-1}$, $S_f = 0{\cdot}0005$ and $d = 0{\cdot}3$ mm) and channel energy slope S_f with changes in width/depth ratio ($Q = 100\ m^3\ s^{-1}$, $Q_s = 0{\cdot}0078\ m^3\ s^{-1}$ and $d = 0{\cdot}3$ mm) using the Meyer-Peter and Müller formula. In most cases there are two possible dynamic equilibrium forms of channel geometry that can transport a given water and sediment load with no net accretion or deposition (A and A'), but only at one point (vertical dashed line) does a unique channel geometry occur. At this point ($Q_{s\,max}$ and $S_{f\,min}$), the total energy expenditure in the flow reaches a minimum, the stationary equilibrium state. (Reprinted with permission from Nanson and Huang (2017).)

$H = 0.3$ indicates a system whereby shear stress of flow is equal to resistance of the channel boundary.

Mass balance equilibrium in a channel requires that Q occurs where there is both an adjustable energy grade slope and *W/D* ratio. In a channel with excess slope (that is, greater than necessary to provide adequate stream power to transport the sediment load), as slope declines an equilibrium channel can have two different values of *W/D* (see Fig. 1.3b). Only at a minimum slope is there a single value of *W/D* that allows Q_s to reach its largest equilibrium value (Fig. 1.3a).

Nanson and Huang recognize that not all channels have excess slope or energy or meet the conditions of adjustable boundaries. They also appreciate that in many streams frequently changing imposed runoff and sediment loads and other environmental changes may make it difficult for equilibrium to be achieved. Their argument is that iterative adjustments nudge the channel toward equilibrium, with selection encouraging and preserving those that move the state closer to $H = 0.3$, with other changes being short-lived due to instability if they occur (Fig. 1.4).

This account is based primarily on Nanson and Huang (2017). Empirical supporting evidence is given in that paper, as well as in a series of other key papers (Huang & Nanson, 2000, 2007; Huang et al., 2004; Nanson & Huang, 2008, 2018).

Fig. 1.4 Channel of Lockyer Creek, Queensland, Australia, in 2013. The creek experienced catastrophic flooding in 2011. As the channel adjusts, according to the Maximum Flow Efficiency theory, changes that move the system closer to mass-balance equilibrium are likely to be preserved, while other changes are unstable and short-lived

1.4.3 Soil Hydrology

In hydrological systems selection operates in the form of contingent partitioning so that ecohydrological systems can adapt to changing moisture inputs. Hydrological dynamics are typically represented by the partitioning of precipitation inputs among various transformations, fluxes, and storages. For instance, the hydroclimatological water balance allocates precipitation inputs to evapotranspiration or runoff outputs and soil moisture storage. Phillips (2023) represented hydrological systems as a partitioning of inputs among outflows and various storage and flux components. Partitioning varies according to the type of hydrological system under investigation and the synoptic context (i.e., wetter, high-input vs. drier, low-input conditions). Stability conditions were determined and linked to two fundamental roles ("jobs") of hydrological systems: The geophysical job is to remove excess water; the ecological job is to enable biogeochemical transfers and transformations necessary for ecosystem functions and to provide adequate water for plants.

Figure 1.5 shows a generalized hydrological system, which was applied to specific cases of soil hydrology, fluvial channel/wetland complexes, and fluviokarst systems, under multiple synoptic scenarios in each case. Stability conditions were determined for various combinations of positive and negative signs for the non-zero

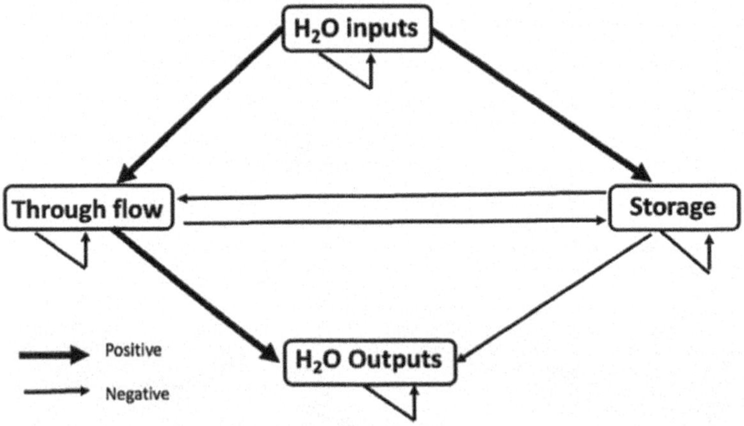

	Inputs	Throughflow	Storage	Outputs
Inputs	a_{11}	a_{12}	a_{13}	0
Throughflow	0	a_{22}	a_{23}	a_{24}
Storage	0	a_{32}	a_{33}	a_{34}
Outputs	0	a_{42}	0	a_{44}

Note: Nonzero entries may be positive, negative, or negligible (≈0) under various circumstances.

Fig. 1.5 Generalized hydrological system. Through flow indicates all water flux through the system, as opposed to saturated throughflow within soil. (Modified from Phillips, 2023)

elements in the interaction matrix shown in Fig. 1.5. The soil example is presented here. As opposed to the stream channel example, in this case stability refers to the dynamical stability of the dynamical system. Stable systems in this sense can return toward prior conditions following a not-too-large change or disturbance. Dynamically unstable systems do not recover toward pre-disturbance conditions.

Several soil hydrology scenarios were examined, based on general principles of rainfall response and soil moisture storage. Figure 1.6 shows the signs of the interaction matrix. The stable scenario is a benchmark case where all signs were selected to achieve dynamical stability. The base case has the same positive and negative links as the box-and-arrow diagram in Fig. 1.5. The dry scenario represents cases with low precipitation inputs, to soil where there is unfilled soil moisture storage

Scenario	a_{11}	a_{22}	a_{33}	a_{44}	a_{12}	a_{13}	a_{23}	a_{24}	a_{32}	a_{34}	a_{42}
Stable benchmark	−	−	−	−	+	+	+	+	−	−	0
1. Base case	−	−	−	−	+	+	−	+	−	−	0
2. Dry	0	0	−	0	+	+	−	+	−	−	0
3. Recharge	−	0	+	0	+	+	−	−	−	−	0
4. Wetting	−	+	−	0	+	+	−	+	−	−	0
5. Saturation	−	−	−	−	+	+	+	+	+	−	−
6. Hortonian	−	−	−	−	+	+	−	+	−	−	−

Note: Links that are always zero are not shown.

Fig. 1.6 Top: Ponding and saturation-excess runoff on a forest trail in western Arkansas. Bottom: Interaction matrix signs for the examined soil hydrology scenarios capacity and no antecedent ponding or flow. In this scenario self-limits other than storage do not apply, and through flow and storage essentially compete for inputs—but storage dominates. The recharge scenario has unfilled soil moisture storage, and wetting front propagation and activation of pores results in self-reinforcement of storage. As storage capacity is approached and self-limits come into play, the wetting scenario occurs. Here flow competes more successfully for the partitioning of inputs, and flow and storage are mutually reinforcing as gravity water and surface ponding and runoff can both feed, and be fed by, flow through the system. All self-limits are operating in the saturation case, and the limits of output capacity may constrain through flow. Infiltration-excess runoff occurs in the Hortonian scenario, as precipitation intensity exceeds infiltration capacity.

Choice, selection, and competition: Inanimate objects such as rocks, channels, water, and air have no desires or intentionality and do not seek, choose, or compete for anything as we understand those terms in the human realm. I use these terms nonetheless because they do reflect what happens, though not necessarily why. Thus "competes more successfully" for flow is a more intuitive and economical way of saying "captures an increasingly larger proportion of the partitioned" flow.

Stability was determined using the Routh-Hurwitz criteria, which can be applied to qualitative, partially-specified systems. The feedbacks, with signs of the a_{ij} omitted, are

$$F_1 = a_{11} + a_{22} + a_{33} + a_{44} \tag{1.5a}$$

$$F_2 = a_{23}a_{32} + a_{24}a_{42} - a_{11}a_{22} - a_{11}a_{33} - a_{11}a_{44} - a_{22}a_{33} - a_{33}a_{44} \tag{1.5b}$$

$$F_3 = a_{23}a_{34}a_{42} - a_{24}a_{42}a_{33} - a_{23}a_{32}a_{44} - a_{22}a_{33}a_{44}. \tag{1.5c}$$

$$F_4 = -\left(a_{23}a_{34}a_{42}a_{11}\right) + a_{23}a_{32}a_{44}a_{11} + a_{24}a_{42}a_{33}a_{11} - a_{11}a_{22}a_{33}a_{44} \tag{1.5d}$$

The Routh-Hurwitz criteria (see Cesari, 1971; Puccia & Levins, 1985) hold that the system is dynamically stable to small perturbations if and only if all $F_i < 0$, and successive Hurwitz determinants are positive. By plugging the positive, negative or zero values indicated in the lower part of Fig. 1.6 into Eqs. (1.5a, 1.5b, 1.5c, and 1.5d), the conditions for stability can be determined. Table 1.1 shows the stability conditions for the various soil hydrology scenarios.

Table 1.1 shows that stability is highly dependent on contingent partitioning of flow, and that the dry and recharge scenarios are generally dynamically unstable, while the soil wetting and saturation cases are generally dynamically stable. The instability in the drier cases allows the soil to adjust to changing conditions (new or additional precipitation inputs) to allow performance of the ecological job. In the wetter situations, dynamical stability allows the system to handle excess moisture inputs (up to a point, anyway) to maintain the general state of the system, thereby fulfilling the geophysical job (Fig. 1.7). Broadly similar results (i.e., dynamical

Table 1.1 Dynamical stability and instability conditions for soil hydrology scenarios

Synoptic Scenario	Stability conditions	Instability conditions	Comments
Benchmark[a]	Unconditionally stable	None	Interaction matrix signs chosen to achieve dynamical stability.
Base case	Self-limits stronger than through flow-storage feedbacks	Through flow-storage feedbacks dominant	Stability contingent on flow-storage feedback strength
Dry[b]	Flow and storage self-limits dominant	Flow-storage partitioning dominant	Stability conditions not plausible in dry scenario.
Recharge[b]	Storage self-reinforcement relatively weak; Self-limits stronger than through flow-storage feedbacks	Storage self-reinforcement relatively strong; Self-limits weaker than flow-storage feedbacks	Instability conditions far more likely.
Soil wetting[a]	Storage, input self-limits stronger than flow self-reinforcement; Self-limits stronger than flow-storage feedbacks	Storage, input self-limits weaker than flow self-reinforcement	Stability likely under wetting conditions.
Soil saturation[a]	Through flow-outflow feedbacks stronger than through flow-storage links	Through flow-outflow feedbacks weaker than through flow-storage links	Instability conditions highly unlikely
Hortonian runoff	Through flow-outflow feedbacks stronger than through flow-storage links.	Through flow-storage feedbacks stronger than through flow-outflow links.	May be stable or unstable, depending on local conditions.

[a]Generally dynamically stable scenarios
[b]Generally dynamically unstable scenarios

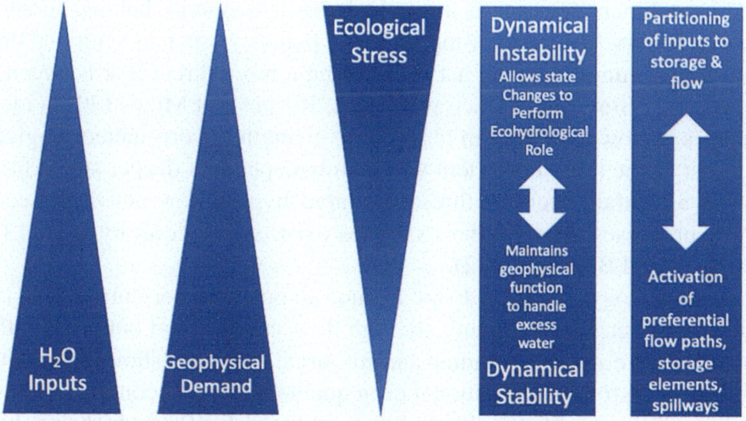

Fig. 1.7 Simplified summary of relationships among water inputs, ecological stress, geophysical demand to move excess moisture, dynamical stability and adaptations. (Reprinted with permission from Phillips (2023).)

instability in dry scenarios; stability in excess-water situations) were obtained for the analyses of channel/wetland complex and fluviokarst systems (Phillips, 2023).

How does selection figure into contingent partitioning and hydrological adaptation? Water balance studies long ago established that soil moisture has "first dibs" on precipitation, generally being recharged before surface runoff or percolation is initiated, at least if infiltration capacity is sufficient to accept the inputs. This is beneficial to vegetation and other biota, and ecohydrological studies have linked hydraulics, ecological filtering, and natural selection, showing that soil systems typically evolve—partly due to ecosystem engineering by plants and other organisms—to optimize soil hydrologic properties serving the ecological role of the soil hydrological system (e.g., Eagleson, 2002; Troch et al., 2013; Nijzink & Schymanski, 2022; Hunt et al., 2023).

Stability often emerges in the wetter soil scenarios as self-reinforcement of flow occurs via gradient and resistance selection. This often happens as preferential flow paths are activated and become connected (e.g., Liu & Lin, 2015; Mohammadi & Illman, 2019; Nieber et al., 2006; Nieber & Sidle, 2010; Sidle et al., 2001; Wilson et al., 2017; Worthington, 2015, 2019). Dynamical stability at saturation is also associated with temporary phenomena such as surface ponding and gravity water which can transition to either flow or storage. Runoff and exfiltration serve as "spillway" mechanisms that allow excess water to be exported, preventing prolonged anaerobiosis and other impacts of waterlogging.

The hydrology literature includes numerous examples of contingent partitioning and switches in flux–storage relationships, though not necessarily framed in terms of dynamical (in)stability, selection, and adaptation. Slattery et al. (2006), for example, identified different modes of runoff generation at the same locations in different precipitation events in an agricultural watershed. Another example is a Swiss watershed found to be unresponsive to precipitation when soil moisture storage is low, but increasingly responsive under higher inputs as storage increases (Teuling et al., 2010). The catchment operates as a simple dynamical system, but less so under dry than wet conditions. With greater inputs, faster pathways such as saturated throughflow and surface runoff become active, creating a more direct link between inputs and outputs. In a Slovenian Flysch watershed, Rusjan and Mikos (2015) identified state changes in flow dynamics in response to changing hydro-meteorological conditions. Most of the time the system was mainly dependent deeper subsurface storage. But at a rainfall intensity threshold, rapid bypass flow activated secondary streamflow processes. Many other examples exist; see reviews by Bonell (1993), Spence (2010) and Blöschl (2022).

The soil hydrology scenarios do not include all potential permutations of interactions and self-effects among inputs, through flow, storage, and output. But the scenarios evaluated are all quite common and are varied enough to show that partitioning of water inputs is strongly conditional even qualitatively, and contingent on synoptic situations. There are 16 possible links (a_{ij}) in Fig. 1.5, 12 of which are always or sometimes nonzero, and nine of which may have at least two different signs (negative, zero and positive). Selection for efficiency in performing the ecological and geophysical jobs occurs in two general ways. One is via system persistence, switching among stable and unstable modes according to geophysical or ecological demand. The second is by positive reinforcement—of preferential flow paths and spillway mechanisms due to gradient and resistance selection, and of soil moisture storage dynamics via ecosystem engineering.

Like selection, adaptation is not restricted to biota and occurs in abiotic aspects of ESS, but it is not omnipresent. Non-adapted hydrological flow systems exist, inevitably so owing to changes and disturbances that delay, disrupt, or prevent development towards adaptation, or simply insufficient time. Evolution of critical traits such as storage capacity, preferential flow paths and spillways is nondeterministic. Rather, these are emergent phenomena subject to selection, and therefore probabilistic and imperfect.

1.5 Summary

Earth systems are strongly influenced by biotic and abiotic factors and processes. The role of selection is generally recognized with respect to biota, in the form of Darwinian natural selection. Biotic-abiotic interactions are evident in ecological filtering and ecosystem engineering. But selection principles also apply to geophysical, abiotic components of ESS.

Natural selection is the best-known example of a wider principle summarized by variation-selection-retention (or recurrence)—the VSR logic. Application of this principle to ESS involves biotic, abiotic, and hybrid phenomena. Abiotic selection has been far less recognized and studied, so that is the focus of this book. In subsequent chapters we review some general ideas and principles related to selection (Chap. 2) and provide an overview of the four major classes of abiotic selection (Chap. 3), followed by chapters devoted to each of these in more detail (gradient, resistance, thermodynamic, and network selection, Chaps. 4, 5, 6, and 7). Chapter 8 explores several examples, both general and specific, of landscape evolution guided by selection. Chapter 9 shows how many phenomena currently or previously interpreted in terms of purported goals of ESS can be more simply explained with VSR logic. The role of selection in integration of reductionist and holistic, bottom-up and top-down approaches is treated in Chap. 10, and the applications of the VSR perspective to planning, engineering, natural hazards, and responses to climate change are explored in Chap. 11.

References

Annila, A., & Salthe, S. (2010). Physical foundations of evolutionary theory. *Journal of Non-Equilibrium Thermodynamics, 35*, 301–321. https://doi.org/10.1515/JNETDY.2010.19

Bejan, A. (2007). Constructal theory of pattern formation. *Hydrology and Earth System Sciences, 11*, 753–768.

Bejan, A. (2017). Evolution in thermodynamics. *Applied Physics Reviews, 4*, 011305. https://doi.org/10.1063/1.4978611

Bejan, A., & Lorente, S. (2013). Constructal law of design and evolution: Physics, biology, technology, and society. *Journal of Applied Physics, 113*, 151301.
Bélanger, J. B. (1849). *Notes sur le cours d'hydraulique, Mémoir, École nat*. Ponts et Chaussés.
Blöschl, G. (2022). Flood generation: Process patterns from the raindrop to the ocean. *Hydrology and Earth System Sciences, 26*, 2469–2480. https://doi.org/10.5194/hess-26-2469-2022
Bonell, M. (1993). Progress in the understanding of runoff generation dynamics in forests. *Journal of Hydrology, 150*, 217–275. https://doi.org/10.1016/0022-1694(93)90112-M
Böss, P. (1919). *Berechnung der Wasserspiegellage beim Wechsel des Fliesszustandes*. Springer.
Cesari, L. (1971). *Asymptotic behavior and stability problems in ordinary differential equations*. Springer.
Corenblit, D., Steiger, J., & Delmotte, S. (2010). Abiotic, residual, and functional components of landforms. *Earth Surface Processes and Landforms, 35*, 1744–1750.
Duniway, M. C., Benson, C., Nauman, T. W., et al. (2022). Geologic, geomorphic, and edaphic underpinnings of dryland ecosystems: Colorado Plateau landscapes in a changing world. *Ecosphere, 13*, e4273. https://doi.org/10.1002/ecs2.4273
Dyke, J. G., Gans, F., & Kleidon, A. (2011). Towards understanding how surface life can affect interior geological processes: A non-equilibrium thermodynamics approach. *Earth System Dynamics, 2*, 139–160.
Eagleson, P. S. (2002). *Ecohydrology: Darwinian expression of vegetation form & function*. Cambridge University Press.
Egbert, M., Hanczyc, M. M., & Harvey, I. (2023). Behavior and the origin of organisms. *Origins of Life and Evolution of Biospheres, 53*, 87–112.
Gerber, E. (1969). Bildung und formen von gratgipfeln und felswänden in den alpen. *Zeitschrift fur Geomorphologie Suppl., 8*, 94–118.
Huang, H. Q., Chang, H. H., & Nanson, G. C. (2004). Minimum energy as the general form of critical flow and maximum flow efficiency and for explaining variations in river channel pattern. *Water Resources Research, 40*, W04502. https://doi.org/10.1029/2003/WR002539
Huang, H. Q., & Nanson, G. C. (2000). Hydraulic geometry and maximum flow efficiency as products of the principle of least action. *Earth Surface Processes and Landforms, 25*, 1–16.
Huang, H. Q., & Nanson, G. C. (2007). Why some alluvial rivers develop an anabranching pattern. *Water Resources Research, 43*, W07441. https://doi.org/10.1029/2006WR005223
Hunt, A. G., Sahimi, M., & Ghanbarian, B. (2023). Predicting streamflow elasticity based on percolation theory and ecological optimality. *AGU Advances, 4*, e2022AV000867. https://doi.org/10.1029/2022AV000867
Kaila, V. R. I., & Annila, A. (2008). Natural selection for least action. *Proceedings of the Royal Society A, 44*, 3055–3070. https://doi.org/10.1098/rspa.2008.0178
Kauffman, S. A. (1971). Cellular homeostasis, epigenesis and replication in randomly aggregated macromolecular systems. *Journal of Cybernetics, 1*, 71–96. https://doi.org/10.1080/01969727108545830
Kauffman, S. A. (1996). *At home in the universe: The search for laws of self-organization and complexity*. Oxford University Press.
Lapenis, A. G. (2002). Directed evolution of the biosphere: Biogeochemical selection or Gaia? *Professional Geographer, 54*, 379–391. https://doi.org/10.1111/0033-0124.00337
Lekevicius, E. (2002). *The origin of ecosystems by means of natural selection*. Lithuanian Academy of Sciences, Institute of Ecology.
Lekevicius, E. (2006). The Russian Paradigm in ecology and evolutionary biology: Pro et contra. *Acta Zoologica Lituanica, 16*, 3–19.
Leopold, L. B. (1994). River morphology as an analog to Darwin's theory of natural selection. *Proceedings of the American Philosophical Society, 138*, 31–47.
Liu, H., & Lin, H. (2015). Frequency and control of subsurface preferential flow: From pedon to catchment scales. *Soil Science Society of America Journal, 79*, 362–377. https://doi.org/10.2136/sssaj2014.08.0330
Lorente, S., & Bejan, A. (2019). Current trends in constructal law and evolutionary design. *Heat Transfer–Asian Research, 48*, 3574–3589. https://doi.org/10.1002/htj.21556

Lorenz, E. N. (1960). Generation of available potential energy and the intensity of the general circulation. In R. F. Pfeffer (Ed.), *Dynamics of climate* (pp. 86–92). Pergamon.

Martyushev, L. M., & Seleznev, V. D. (2006). Maximum entropy production principle in physics, chemistry and biology. *Physics Reports, 426*, 1–45.

Mohammadi, Z., & Illman, W. A. (2019). Detection of karst conduit patterns via hydraulic tomography: A synthetic inverse modeling study. *Journal of Hydrology, 572*, 131–147. https://doi.org/10.1016/j.jhydrol.2019.02.044

Murray, A. B., Knaapen, M. A. F., Tal, M., & Kirwan, M. L. (2008). Biomorphodynamics: Physical-biological feedbacks that shape landscapes. *Water Resources Research, 44*, W11301.

Nanson, G. C., & Huang, H. Q. (2008). Least action principle, equilibrium states, iterative adjustment and the stability of alluvial channels. *Earth Surface Processes and Landforms, 33*, 923–942.

Nanson, G. C., & Huang, H. Q. (2017). Self-adjustment in rivers: Evidence for least action as the primary control of alluvial-channel form and process. *Earth Surface Processes and Landforms, 42*, 575–594.

Nanson, G. C., & Huang, H. Q. (2018). A philosophy of rivers: Equilibrium states, channel evolution, teleomatic change and the least action principle. *Geomorphology, 302*, 3–19.

Nieber, J. L., & Sidle, R. C. (2010). How do disconnected macropores in sloping soils facilitate preferential flow? *Hydrological Processes, 24*, 1582–1594. https://doi.org/10.1002/hyp.7633

Nieber, J. L., Steenhuis, T. S., Walter, T., & Bakker, M. (2006). Enhancement of seepage and lateral preferential flow by biopores on hillslopes. *Biologia, 61*, 225–228. https://doi.org/10.2478/s11756-006-0162-1

Nijzink, R. C., & Schymanski, S. J. (2022). Vegetation optimality explains the convergence of catchments on the Budyko curve. *Hydrology and Earth System Sciences, 26*, 6289–6309.

Ozawa, H., Ohmura, A., Lorenz, R. D., & Pujol, T. (2003). The second law of thermodynamics and the global climate system: A review of the maximum entropy production principle. *Reviews of Geophysics, 41*. https://doi.org/10.1029/2002RG000113

Paltridge, G. W. (1975). Global dynamics and climate—A system of minimum entropy exchange. *Quarterly Journal of the Royal Meteorological Society, 101*, 475–484.

Paltridge, G. W. (1978). The steady-state format of global climate. *Quarterly Journal of the Royal Meteorological Society, 104*, 927–945.

Paltridge, G. W. (1981). Thermodynamic dissipation and the global climate system. *Quarterly Journal of the Royal Meteorological Society, 104*, 531–547.

Pascal, R., & Pross, A. (2016). The logic of life. *Origins of Life and Evolution of Biospheres, 46*, 507–513.

Phillips, J. D. (2008). Goal functions in ecosystem and biosphere evolution. *Progress in Physical Geography, 32*, 51–64. https://doi.org/10.1177/0309133308089497

Phillips, J. D. (2009). Biological energy in landscape evolution. *American Journal of Science, 309*, 271–290.

Phillips, J. D. (2021). *Landscape evolution. Landforms, ecosystems, and soils*. Elsevier.

Phillips, J. D. (2023). Contingent partitioning and adaptation in hydrological systems. *Ecohydrology, 16*, e2567. https://doi.org/10.1002/eco.2567

Pross, A., & Khodorkovsky, V. (2004). Extending the concept of kinetic stability: Toward a paradigm for life. *Journal of Physical Organic Chemistry, 17*, 312–316.

Pross, A., & Pascal, R. (2017). How and why kinetics, thermodynamics, and chemistry induce the logic of biological selection. *Bielstein Journal of Organic Chemistry, 13*, 665–674.

Puccia, C. J., & Levins, R. (1985). *Qualitative modeling of complex systems*. Harvard University Press.

Rusjan, S., & Mikos, M. (2015). A catchment as a simple dynamical system: Characterization by the streamflow component approach. *Journal of Hydrology, 527*, 794–808. https://doi.org/10.1016/j.jhydrol.2015.05.050

Scheidegger, A. E. (1987). The fundamental principles of landscape evolution. *Catena Supplement, 10*, 199–210.

Scheidegger, A. E. (1991). *Theoretical geomorphology* (3rd ed.). Springer.

Sharma, V., & Annila, A. (2007). Natural process—Natural selection. *Biophysical Chemistry, 127*, 123–128.
Sidle, R. C., Noguchi, S., Tsuboyama, Y., & Laursen, K. (2001). A conceptual model of preferential flow systems in forested hillslopes: Evidence of self-organization. *Hydrological Processes, 15*, 1675–1692. https://doi.org/10.1002/hyp.233
Slattery, M. C., Gares, P. A., & Phillips, J. D. (2006). Multiple modes of run-off generation in a North Carolina coastal plain watershed. *Hydrological Processes, 20*, 2953–2969. https://doi.org/10.1002/hyp.6144
Smith, C. H. (1986). A contribution to the geographical interpretation of biological change. *Acta Biotheoretica 35*(4), 229e278. https://doi.org/10.1007/BF00047097
Spence, C. (2010). A paradigm shift in hydrology: Storage thresholds across scales influence catchment runoff generation. *Geography Compass, 4*, 819–833. https://doi.org/10.1111/j.1749-8198.2010.00341.x
Strömberg, C. A. E. (2011). Evolution of grasslands and grassland ecosystems. *Annual Review of Earth and Planetary Sciences, 39*, 517–544.
Teuling, A. J., Lehner, I., Kirchner, J. W., & Seneviratne, S. I. (2010). Catchments as simple dynamical systems: Experience from a Swiss pre-alpine catchment. *Water Resources Research, 46*, e2009WR008777. https://doi.org/10.1029/2009WR008777
Troch, P. A., Carrillo, G., Sivapalan, M., et al. (2013). Climate-vegetation-soil interactions and long-term hydrologic partitioning: Signatures of catchment co-evolution. *Hydrology and Earth System Sciences, 17*, 2209–2217.
Vernadsky, V. I. (1926). *The biosphere*. 1998 edition translated from Russian by D. B. Langmuir, edited by M. A. S. McMenamin & L. Margulis. Nevramont.
Vitas, M., & Dobovisek, A. (2019). Towards a general definition of life. *Origins of Life and Evolution of Biospheres, 49*, 77–88.
Wellborn, G. A., & Langerhans, R. B. (2015). Ecological opportunity and the adaptive diversification of lineages. *Ecology and Evolution, 5*, 176–195.
Wilson, G. V., Nieber, J. L., Fox, G. A., et al. (2017). Hydrologic connectivity and threshold behavior of hillslopes with fragipans and soil pipe networks. *Hydrological Processes, 31*, 2477–2496. https://doi.org/10.1002/hyp.11212
Wong, M. L., Cleland, C. E., Arend, D., Jr., et al. (2023). On the roles of function and selection in evolving systems. *Proceedings of the National Academy of Sciences (USA), 102*, e2320223120.
Worthington, S. R. H. (2015). Characteristics of channel networks in unconfined carbonate aquifers. *Geological Society of America Bulletin, 127*, 759–769. https://doi.org/10.1130/B31098.1
Worthington, S. R. H. (2019). How preferential flow delivers pre-event groundwater rapidly to streams. *Hydrological Processes, 33*, 2373–2380. https://doi.org/10.1002/hyp.13520

2 Key Principles

Abstract

Several overarching principles are applicable to selection in Earth systems. (1) The generalized selection logic of VSR (variation-selection-retention) that applies to abiotic selection in Earth surface systems (ESS), as well as to biology. Darwinian natural selection is a version of the broader VSR principle. (2) Positive feedback is common in ESS and is in many cases a form of selection. While positive feedback sometimes produces instability, it can also serve to reinforce variations or disturbances that are beneficial for system function and survival. (3) Abiotic selection repeatedly produces similar patterns as emergent phenomena, which is a simpler explanation than those that imply intentionality on the part of inanimate nature. (4) Selection is probabilistic, not deterministic. On average and in the aggregate, more efficient and better adapted features persist and recur, but this is not always the case (just as the fitter organisms are not the ones to survive and reproduce in every instance). (5) The first and second laws of thermodynamics (conservation of matter, energy, and momentum) are involved in selection, as reflected in the least action and maximum entropy production principles. (6) Gradient selection involves preferential utilization of the steepest energy gradient pathways, often imprinted and reinforced by positive feedback. (7) Resistance selection is based on fundamental principles of force vs. resistance, resulting in increased survival of stronger, more durable, and more stable entities. (8) Gradient and resistance selection are part of a broader class of efficiency selection, to which can be added network selection and thermodynamic selection.

J. D. Phillips, *Abiotic Selection in Earth Surface Systems*, Geophysics and Environmental Physics, https://doi.org/10.1007/978-3-031-85862-8_2

2.1 Necessary and Sufficient Conditions for Evolution

2.1.1 VSR and Generalized Selection

In Chap. 1 we encountered the VSR logic: *Variations* occur; advantageous ones are *Selected*; the latter are preferentially *Retained*. This encapsulates the necessary and sufficient conditions for evolution by selection to occur. In natural selection the variations are organismal traits, selection is based on fitness for habitats and reproductive success, and retention occurs via propagation of the advantageous traits in the gene pool.

In his book *Darwin's Dangerous Idea,* Dennett (1995) characterizes natural selection as a general algorithm involving variation, selection and retention that is independent of context and applicable well beyond biology. A key component of his argument is that a VSR framework can explain many things that are otherwise difficult to explain without invoking a guiding hand.

*Generalized (*or *Universal) Darwinism* holds that the principles of variation, selection, and retention (preservation and replication) that are the cornerstone of Darwinian evolution are applicable to darn near anything. Though present in many science and social science fields, generalized Darwinism is most prominent in evolutionary economics. In many cases generalized Darwinism depends on, or is at least framed in terms of, analogies with biological evolution. Such analogies are unnecessary in Earth surface systems (ESS), as selection and VSR logic apply directly to geophysical and other environmental phenomena.

Vellend's (2016) theory of ecological communities also recognizes selection as a more general process beyond Darwinian natural selection, as does the constructal law for flow systems mentioned in Chap. 1. Lorente and Bejan (2019) explicitly addressed the constructal law as an evolutionary selection process, the basic principles of which apply to thermodynamic systems in general. The review of maximum entropy production theories by Bruers and Meysman (2008) characterizes non-variational maximum entropy theories as selection principles.

Self-organization of complex structures is thermodynamically favored, Ueltzhöffer et al. (2021) argued, whenever those structures dissipate free energy that could not be accessed otherwise. This also applies to relatively simple dissipative systems such as convection cells, hurricanes, lightning strikes, or mechanical cracks (as well as to complex biological systems). Environmental dynamics can be effectively represented via links to the thermodynamic efficiency of underlying physical processes. Ueltzhöffer et al. (2021) demonstrated a simple mechanism that leads to self-selection of efficient dissipative structures in a stochastic chemical reaction network, when the dissipated driving chemical potential difference is decreased. They also showed how such a trend can emerge in a hierarchy of self-similar dissipative structures, each of which feeds upon dissipation from a previous level, when moving away from the initial, driving disequilibrium.

Recapitulating and synthesizing many of these ideas, Wong et al. (2023) stated that evolving systems appear to be conceptually equivalent in that they display three notable attributes: (1) They form from numerous components that have the potential to adopt combinatorially vast numbers of different configurations; (2) processes exist that generate numerous different configurations; and (3) configurations are preferentially selected based on function. They identified universal concepts of selection—static persistence, dynamic persistence, and novelty generation—that underpin function and drive systems to evolve through the exchange of information between the environment and the system.

2.2 Positive Feedback as Selection

Negative feedbacks in ESS, and in systems in general, are responses to changes or disturbances that function to maintain the system state or at least return the system toward its pre-disturbance condition—a thermostat is a classic analog. Positive feedback, by contrast, operates to exaggerate or amplify changes or perturbations, causing them to persist or grow disproportionately relative to the driving factor or event. Ice-albedo feedback in the ocean-atmosphere system is a canonical case. Higher temperatures result in reduced sea-ice, reducing albedo of the sea surface, resulting in more solar radiation absorption, leading to further warming. Cooling that leads to increased ice cover raises albedo, reflecting more incoming solar radiation, reinforcing the cooling trend.

In the literature on biological evolution and natural selection, the role of positive and negative feedbacks resulting from natural selection is not always distinguished from selection itself as feedback. Robertson's (1991) review of feedback in Darwinian evolution characterizes positive feedback as producing "instability and even catastrophe," a view that persists today (e.g., Pausas & Bond, 2022). This is indeed often, but not always, the case. At the system scale, dynamical stability is contingent on *net* negative feedback, but systems with only negative feedback links are unstable. Further, dynamical instability where positive feedbacks drive development of a new state or configuration are often advantageous or necessary for adaptation; Robertson (1991) and Crespi (2004) give several examples. Indeed, an oft-cited biological evolution phenomenon is the so-called "arms race." For example, predators develop more lethal claws and fangs, and prey species respond by evolving thicker skin and armor (Vermeij, 1987 is a classic reference on this).

Evolutionary biologists and ecologists have often focused on negative feedback dominated stable equilibrium systems (Crespi, 2004). This is no doubt partly attributable to the long dominance of balance-of-nature perspectives, but Crespi also points out that positive feedback dynamics are often fleeting at evolutionary time scales, making them less likely to be observed. Certainly, no amplification or escalation can continue indefinitely, as eventually laws of physics or resource limitations constrain growth.

Organisms modify the abiotic environment, and ecosystem engineering and niche construction are often positive (that is, benefiting the engineer organism). As positive ecosystem engineering improves habitat for the engineer, and positive niche construction conveys evolutionary advantages to the constructor, these effects link positive feedback to natural and ecological selection.

In ESS and geophysical processes most, if not all, positive feedback trends meet the first two VSR criteria—they involve variability in time and/or space, which are selected by positive feedbacks to be enhanced or amplified. When these are retained, the third criterion is also met. In biology, directionality in history emerges from selection in the form of positive feedbacks between resources and consumers (Vermeij, 2011).

Many examples will be given later, but a quick, simplified one here: Concentrated water flows on the land surface (representing variations in force or shear stress) are sometimes sufficient to cause excavation of a channel. The channel, providing a more advantageous flow path and drawing water from its adjacent areas, attracts more water. The increased water, plus greater flow depths than nearby alternative pathways, further increases shear stress, promoting further enlargement of the channel. This can continue until the supply of runoff, or the geotechnical stability of the channel banks become limiting, after which the channel commonly persists.

Selection in geophysical phenomena cannot be characterized in terms of goals, as abiotic entities have no intentionality. Because selection is probabilistic, it also cannot be characterized in terms of universal deterministic laws. Rather, the outcomes of selection (e.g., soil hydrological systems, global temperature and circulation patterns, channel networks) are emergent properties, as outlined in detail in Chap. 9. Recurrence of some features or patterns understandably leads to attempts to interpret them as outcomes of some law of nature or goal function of ESS. But try this thought experiment: do any laws dictate that in areas of layered sedimentary rocks, sandstone ridgetops (Fig. 2.1) must form? Does nature somehow prefer sandstone ridgetops? Simpler, and more logical in terms of not postulating hidden laws or attributing goals to inanimate features, is the explanation that sandstones are more physically durable and weathering resistant than the lithologies they often

Fig. 2.1 The author amid sandstone ridgetops on the dissected plateau of the Blue Mountains, New South Wales, Australia

occur alongside in sedimentary sequences. As denudation proceeds, they are *selectively* preserved as weathering and erosion *selectively* remove less resistant material. As a result, tablelands and plateaus in sedimentary rocks around the world have sandstone ridgetops. This emergent, selection-based explanation is simpler than any law or goal-function based account, and therefore preferred by Ockham's Razor.

Emergence vis selection is simpler than attempts to ascertain and explain why rivers, atmospheric energy transfers, biogeochemical cycles, or chemical reactions (for instance), should seek or prefer anything.

2.3 Selection Is *Probabilistic* and *Local*

Natural selection is probabilistic, not deterministic. On average, fitter and better-adapted organisms thrive, survive, and reproduce more often than less fit and more poorly adapted relatives. However, even the strongest, fastest, and smartest fish might be caught by a predator, while slower, weaker, and dumber ones escape through sheer good fortune.

So it is also with geophysical and abiotic selection. The most efficient, stable, and durable features and processes have a much higher probability of occurrence and persistence, but this is not inevitable or assured. Thus, efficiency selection in geomorphic and hydrologic systems probabilistically nudges them toward "fitness," but does not predetermine any particular outcome (e.g., Khatibi, 2011; Nanson & Huang, 2018; Phillips, 2021: Ch. 8).

Natural selection operates on individuals (though it may be manifested at much broader levels) and is thus localized. The same goes for abiotic selection, as fluxes, transformations, and storage of energy, matter, and information in ESS are generally only directly affected by the gradients and materials at hand locally. This, while there might be globally more efficient ways to redistribute heat or deplete pressure gradients (for instance), air parcels can only respond to energy balances and pressure gradients in the immediate vicinity.

2.4 Energy and Mass

Many general principles and laws apply to geophysical, geochemical, and biological processes. However, only a few are applicable everywhere and always, regardless of temporal or spatial context.

The first law of thermodynamics is the law of energy conservation, which holds that energy cannot be created or destroyed, only changed in form. The second law states that no energy conversion is completely efficient; some (thermodynamic) entropy is always produced. This also means that the total entropy of interacting systems never decreases. A common corollary is that heat spontaneously flows from hotter to colder bodies or locations. This can be generalized to a more general principle of gradient fluxes, whereby energy and mass move along gradients from higher toward lower concentrations or potentials.

Two other thermodynamic laws exist, but they more rarely come into play in analysis of ESS. The "zeroth" law defines thermal equilibrium: If two systems are in thermal equilibrium with a third system, they are also in thermal equilibrium with each other. The fourth law also pertains to temperature and is applicable mainly to conditions where temperatures are lower than those encountered in the Earth-atmosphere system: As the temperature approaches zero, a system's entropy approaches a constant value.

The first and second laws explain, in a nutshell, why the least action principle (see below) is consistent with, and perhaps tantamount to, the principle of maximum entropy production. If the work accomplished by a given amount of energy is performed as efficiently as possible (i.e., using the minimum amount of the available energy), there will often be some energy left over. As energy cannot be destroyed, it must be dissipated as entropy (typically in the form of heat and friction).

Matter, measured by mass, represents stored energy; their equivalence established by Einstein's famous $E = MC^2$ (energy equals mass times the speed of light squared). Conservation of mass is required in any chemical reaction, for instance—since mass is neither created nor destroyed, the total mass of reactants equals the total mass of products. When mass is converted to energy, the total mass/energy is conserved.

Conservation laws also apply to momentum, a product of an object's mass (M) and velocity v: momentum = Mv. The momentum of a collection of objects does not change unless the system is acted on by external forces. This applies to both linear momentum (a vector quantity with both magnitude and direction) and angular momentum. The latter applies to rotational motion. Conservation of angular momentum explains, for example, why a tropical cyclone which slows its forward movement may increase the size of the rotating circulation to compensate.

These laws come into play in Bernoulli's principle in fluid mechanics, which states that increases of the speed of a fluid occur simultaneously with decreases in static pressure and/or the fluid's potential energy. That is, the total energy remains constant along a streamline. Bernoulli's equation is

$$p_1 + \frac{\rho {v_1}^2}{2} + \rho g y_1 = p_2 + \frac{\rho {v_2}^2}{2} + \rho g y_2 \tag{2.1}$$

where p is pressure, ρ is the density of the fluid, v is velocity, g the gravity constant (9.81 m sec^{-2}), and y the height above ground. The subscripts indicate two locations along the flow line. Bernoulli's equation can also be expressed as

$$p_1 + \frac{\rho v^2}{2} + \rho g y = constant \tag{2.2}$$

The Bernoulli equation applies to incompressible fluids in the absence of friction. The conservation laws also give us the ideal gas law,

$$pV = nRT \tag{2.3}$$

where n is the number of moles of an ideal gas, R is a constant, and T = temperature. While no real-world gas is truly ideal (in an ideal gas the particles do not attract or repel each other, and the gas takes up no space), the law provides a good approximation of real gases in most situations relevant to geosciences, and reflects the critical idea that if temperature, pressure, or volume changes, at least one of the others must also change to maintain the conservation of mass and energy. Thus, for example, as rising air experiences lower pressure from the atmosphere above, it expands (increases volume), and its temperature decreases.

2.4.1 Least Action Principle

To reiterate from Chap. 1, in mechanics the least action principle (LAP) states that in a conservative dynamical system the motion between any two points is such that the action has a minimum value with respect to all paths between the points that correspond to the same energy. Informally, "nature always finds the most efficient path;" formally, for a moving object the average kinetic energy minus the average potential energy is as little as possible for the path of an object going from one point to another. The LAP is manifested in geophysics as the performance of work with as little energy as possible (i.e., maximizing efficiency). Given the necessary information, for the case of say, a cannonball or a fly ball or a space capsule re-entering the atmosphere, a path of motion prescribed by the LAP can be calculated straightforwardly. In ESS the situation is almost always more complicated, involving multiple masses or objects and varying resistances and other (than the one driving the motion) forces along the way. Furthermore, as opposed to a baseball or a meteorite flying through the air, flowing water or boundary layer winds, for instance, may face barriers to following the most energetically efficient path. A tumbling rock or an air parcel in a valley cannot "see" alternative paths available to it; it is only affected by the gradients and potentials in the immediate vicinity. ESS, though subject to the LAP with respect to mechanics at the microscale, are characterized by multiple processes and elements, different selective mechanisms, multiple degrees of freedom, and interactions among their elements.

The first to connect the LAP with selection in ESS were He Huang and Gerald Nanson, as described in Chap. 1. Kaila and Annila (2008) juxtaposed the LAP and Darwinian natural selection in these terms:

Take the shortest path ⟷ *Take the fittest unit*

They further argued that the two principles reconcile to a single principle, whereby evolution of all types is a "non-Euclidean energy density landscape in flattening motion."

Starting with Sharma and Annila's (2007) expression of the second law of thermodynamics as a differential equation of motion for the probability P,

$$\frac{dP}{dt} = LP, \tag{2.4}$$

where L is a propagator

$$L = \frac{1}{k_B T} \sum_{j,k} \frac{\partial x_k}{\partial x_j} \left[\frac{\partial \mu_k}{\partial x_j} - \frac{\partial Q_k}{\partial x_j} \right] \tag{2.5}$$

L drives the transport ($dx_k/dt = -\sum dx_j/dt$) as, e.g., currents and diffusion between general coordinates x_k, x_j by drawing down potential energy gradients $\partial\mu_k/\partial x_j$ and the fields coupling to the j,k transport process $\partial Q_k/\partial x_j$. $k_B T$ is the average energy per particle.

Equation (2.5) has an alternate form where the driving forces and gradients are chemical potentials (Sharma & Annila, 2007; Kaila & Annila, 2008). Equation (2.5) is the appropriate form when distinct components or entities in the system are determined by a spatial gradient.

By the second law, $dH/dt \geq 0$, where H is thermodynamic entropy. In terms of the logarithmic probability $dH = k_B\, d(\ln P)$, the equation of motion for the resolved transformation is

$$\frac{dH}{dt} = \frac{k_B}{P}\frac{dP}{dt} = k_B L = \frac{1}{T} \sum_{j,k} v_k \left(\frac{\partial \mu_k}{\partial x_j} - \frac{\partial Q_k}{\partial x_j} \right) \geq 0, \tag{2.6}$$

where $v_k = dx_k/dt$ is the flow. Again, there is an analogous equation of the same form as Eq. (2.6) for chemical gradients and potentials.

If the potential μ_k at x_k is higher than μ_j at x_j, then the gradient and the flux will be negative. That is, $\partial\mu_k/\partial x_j - \partial Q_k/\partial x_j < 0$, and $v_k > 0$ (flow will be positive and the inequalities flipped if the potential differences are reversed). Then $dH/dt > 0$ until the gradients are depleted and $dH/dt = 0$.

Kaila and Annila (2008) then develop the differential equation of motion describing the increase in entropy and decrease in free energy due to flows along potential gradients:

$$\frac{d}{dT} TS = -\sum_{j,k} v_j \left(\frac{\partial \mu_k}{\partial x_j} - \frac{\partial Q_k}{\partial x_j} \right) = -\sum_{j,k} D_j \left(< \mu_k - Q_k \right) \tag{2.7}$$

Equation (2.7) indicates a flow of kinetic energy from x_k to x_j is due to the decreasing potential μ_k, simultaneous with the dissipation $\partial Q/\partial t = -\sum v_j\, \partial Q_k/\partial x_j$.

Flows occur from high to low potentials along the steepest gradients of energy for both $\partial\mu_k/\partial x_j$ and $\partial Q_k/\partial x_j$. Kaila and Annila (2008) go on to develop an integral form of Eq. (2.7) to show that the steepest gradients also correspond to the shortest paths. Key points that will be revisited later in the book are *gradient depletion,* whereby not only are gradients depleted by high-to-low fluxes, but fluxes that do so as rapidly as possible are selected for.

2.5 Optimality and Selection

Earth and environmental sciences feature several explanations based on the idea that development of ESS is characterized by tendencies to maximize or minimize some aspect of energy or mass flux or organizational characteristics. These extrema often increase stability or efficiency of ESS, and for brevity are referred to here as optimal principles, whether they explicitly use that term or not. As many of these propositions are based on maximizing or minimizing something, they have also been referred to as extremal hypotheses. Most of these are effective, at least to some extent and within certain domains, in accounting for empirical observations. But abiotic and ecological phenomena lack intentionality, and we are hard put to explain why an ESS should optimize, minimize, or maximize anything. Rocks and rivers don't care, and while the individuals within it might care, in some sense of the word, ecosystems don't give a damn either (see, e.g., Nanson & Huang, 2008; Phillips, 2011, 2021; Quijano & Lin, 2014).

Optimal principles have been proposed in fluid dynamics, climatology, hydrology, geomorphology, and ecology. Though they differ in terminology and in the aspects of ESS addressed, many of these optimality principles are consistent with each other, and not uncommonly equivalent with respect to the bottom-line implications or predictions. For example, Ozawa et al. (2003) showed the equivalency of optimal principles related to atmospheric heat flux, global climate, fluid convection, and turbulent dissipation. Many extremal principles related to hydraulic geometry (the mutual adjustments of channels and the flows they convey) have been shown to be equivalent with respect to their fundamental hydrological and geomorphological implications (Griffiths, 1984; Phillips, 1990, 1991; Huang & Nanson, 2000; Nanson & Huang, 2008; Paik & Kumar, 2010). Most optimal principles of ecosystem development are mutually consistent and even formally related, and the metrics are highly correlated (Patten, 1995; Fath et al., 2001; Yen et al., 2014).

Some optimal principles are directly linked to the LAP by proposing maximum efficiency in energy use and/or mass fluxes. Others are either directly based on maximum entropy production (MEP) or propose maximum energy throughput, which also implies MEP (Fath et al., 2001; Ozawa et al., 2003; Dewar, 2005; Kleidon et al., 2010). These are linked to thermodynamic selection, and explored further in Chap. 9.

2.6 Gradient Selection and Self-reinforcement

Water flows downhill. Water follows the path of least resistance. These common-sense observations are consistent with gradient selection.

For instance, consider the Darcy-Weisbach equation, originally developed as an empirical expression for head loss in flow along a pipe, adapted to runoff and other flows:

$$V = \left(8g\,RS/f\right)^{0.5} \tag{2.8}$$

where g is the gravity constant, V is velocity, R is hydraulic radius, S is the energy gradient slope, and f is a friction factor. R is cross-sectional area divided by wetted perimeter, and for many surface and channel flows is well approximated by mean depth d. S is often approximated, for practical reasons, by water surface or channel bed slope. The other flow resistance equations commonly in use, the *Chézy* and Manning equations, are similar in form:

$$V = C(RS)^{0.6} \tag{2.9}$$

$$V = k / n R^{2/3} S^{1/2} \tag{2.10}$$

The empirical Chézy coefficient C incorporates roughness or friction as well as the gravity constant, n is the empirical Manning roughness coefficient, and $k = 1$ when SI units are used.

Taking velocity as the indicator of flow efficiency to be selected for, it is clear (and consistent with intuition and experience) that water will preferentially follow flow paths that are deeper, steeper, and smoother (less roughness or friction). Locally, the steepest energy gradient slopes lead into greater depths, and for a given surface or boundary, greater depths are associated with lower roughness as roughness elements are submerged. The flow equations are therefore fully consistent with a broadly phrased principle of gradient selection, where steeper gradients and lower resistance are associated with higher velocities and maximum flow efficiency.

The manifestation of gradient selection at Earth's surface is strongly linked to positive feedbacks. The mean boundary shear stress τ of water flows is given by.

$$\tau = \gamma RS. \tag{2.11}$$

The specific gravity of water (γ) is the product of density and the gravity constant ($\gamma = \rho_\omega g$). Cross-sectional stream power is

$$\Omega = \gamma QS = \tau WV, \tag{2.12}$$

where discharge Q is the product of velocity and cross-sectional area A, itself the production of flow width and mean depth: $Q = A\ V = w\ d\ V$. Power per unit weight of water reduces to the product VS.

The selection of deeper, steeper, faster flow paths therefore increases the shear stress and the stream power. The shear stress often exceeds the thresholds for boundary erosion and channel incision, and sediment transport capacity is proportional to stream power. Erosion and transport tend to develop and (to a finite limit) enlarge channels and remove debris that would otherwise increase roughness. This makes the flow path even more efficient (and draws in more water due to local slope gradients and water table drawdown). Preferential flow paths, channels and channel networks develop, initially, as a product of gradient selection.

This hydraulic example is but one instance of gradient selection in ESS and the importance of positive feedback in reinforcing flow pathways. In Chap. 4, additional examples will be explored, including atmosphere flows, hillslope mass movements, aeolian processes, karst conduit development, and subglacial stream capture.

2.6.1 Force vs. Resistance

The principle of resistance selection is a form of selection-by-persistence: More resistant materials and forms last longer than weaker, less resistant ones. Resistance is context sensitive, depending on the type and magnitude of force applied. Many limestones, for instance, are resistant to physical erosion by water, but have low resistance to chemical dissolution. Physical rock strength itself has multiple aspects, including compressive, tensile, shear, and impact strength.

Force also has multiple aspects, even when the same process at the same time and place is considered. For instance, wind force per unit of cross-sectional area A exposed to the wind varies as the square of wind velocity, and wind load F (in kg) is given by:

$$F = 0.613V^2 AC_D \tag{2.13}$$

C_D is a drag coefficient. The wind load is also affected by dynamic pressure and the mass of air and its contents, such as suspended sediment (throw weight). Wind loads reflect the cumulative impacts of constantly applied pressure, additional pulsing of shorter wind gusts, occasional rolling shock waves of high pressure, acceleration and deceleration around a mean value, and variable throw weights of the wind stream. The full dynamic nature of winds cannot be fully represented by mean or peak velocities. With trees, for example, the magnitude and periodicity of swaying, along with the frequency of wind pressure peaks, can generate tremendous synergies of load and resistance (Coder, 2018). Failure due to material fatigue is affected by stress amplitude, but also by the number of cycles (e.g., by wind calm/gust sequences).

Force/resistance analyses in Earth sciences are sometimes based on a strict Newtonian physics definition of force, and sometimes on a broader concept of the processes or factors that drive change, such as the pH of water as a force of dissolution versus the solubility of material as resistance.

A common example of applied force/resistance is employed in engineering geology to assess the stability of hillslopes, the F-ratio, with F representing a factor of safety. It is the ratio of shear stress of materials on sloping surfaces versus the shear strength. This will be explored in more detail in Chap. 5.

2.7 Efficiency Selection

Gradient selection works to enhance flows and matter-energy fluxes and can thus be considered as a class of the broader category of efficiency selection. In some cases, resistance selection is the converse of gradient selection, and is at least related to, if not a subset of, efficiency selection. Other types of efficiency selection are network and thermodynamic selection, covered in Chaps. 6 and 7.

2.7.1 Network Selection

Network selection occurs as some flow networks or patterns of interconnection within ESS are more effective in supporting system functions and survival. When these networks or patterns begin to form or emerge, they are thus selected for. Perhaps the canonical example in flow networks is the recurrence of branching, dendritic channel patterns in fluvial systems in a wide variety of settings. The branching dendritic pattern is also selected for in a variety of biological systems (e.g. circulatory systems, plant root and branching architectures), and where they can form, these topologies are the most efficient way to gather flows from an area and deliver them to a single location (e.g., the mouth of a watershed), deliver flows from a central location throughout an area (e.g. distribution networks), or both collect and distribute flows (e.g., circulatory systems). Other examples include selective reinforcement of networks that facilitate, e.g., more rapid biogeochemical cycling or multiple options for hydrologic partitioning.

Applications of percolation theory (see Chap. 6) to a variety of Earth system features and processes shows, repeatedly, that pathways for water flow, solute transport, root growth, and other processes are characterized by high- or maximum-efficiency networks.

With respect to interactions among components of ESS, selection often favors higher connectedness and denser interaction patterns. While these are not advantageous in every case (think of the spread of viruses of all types), in many environmental systems high connectivity enhances efficiency and convers other advantages, which typically outweigh any drawbacks. Thus, selection often encourages increasing connectivity.

2.7.2 Thermodynamic Selection

The conservation laws for energy, mass, and momentum are the most fundamental physical laws and constraints on Earth. Thermodynamic selection, as described in Chap. 2 and treated in more detail in Chap. 7, involves fluxes and transformations, constrained by the conservation laws, along various literal and figurative pathways. From among these, those that are most effective in depleting gradients and in performing work most efficiently tend to persist and recur, while others are more ephemeral.

Thermodynamic selection in fluid dynamics is manifested in atmospheric and oceanic circulation patterns, climate phenomena involved in tropics-to-polar heat redistribution, and aeolian processes, to name a few. Thermodynamic selection also plays a role in biogeochemical selection and is often closely linked to network selection.

In the next chapter an overview of various forms of selection in or related to geophysics is given, including the role of abiotic factors in natural selection and ecological filtering, and impact of biological evolution on the geological environment.

References

Bruers, S., & Meysman, F. (2008). *A useful correspondence between fluid convection and ecosystem operation*. arXiv physics-fluid dynamics, arXiv:0708.0091.

Coder, K. D. (2018). *Trees and storm wind loads*. Warnell School of Forestry & Natural Resources, University of Georgia, Thompson Mills Forest & State Arboretum Outreach Product ARBORETUM-18-F 48 p.

Crespi, B. J. (2004). Vicious circles: Positive feedback in major evolutionary and ecological transitions. *Trends in Ecology and Evolution, 19*, 627–633. https://doi.org/10.1016/j.tree.2004.10.001

Dewar R. C. 2005. Maximum entropy production and non-equilibrium statistical mechanics. In Kleidon, A., Lorenz, R.D. (eds), Non-Equilibrium Thermodynamics and the Production of Entropy. Understanding Complex Systems. *Berlin: Springer*, p. 41–55.

Dennett, D. C. (1995). *Darwin's dangerous idea*. Simon and Schuster.

Fath, B. D., Patten, B. C., & Choi, J. S. (2001). Complementarity of ecological goal functions. *Journal of Theoretical Biology, 208*, 493–506.

Griffiths, G. A. (1984). Extremal hypotheses for river regime: An illusion of progress. *Water Resources Research, 20*, 113–118.

Huang, H. Q., & Nanson, G. C. (2000). Hydraulic geometry and maximum flow efficiency as products of the principle of least action. *Earth Surface Processes and Landforms, 25*, 1–16.

Kaila, V. R. I., & Annila, A. (2008). Natural selection for least action. *Proceedings of the Royal Society A, 44*, 3055–3070. https://doi.org/10.1098/rspa.2008.0178

Khatibi, R. (2011). Evolutionary systemic modeling of practices on flood risk. *Journal of Hydrology, 401*, 36–52.

Kleidon, A., Malhi, Y., & Cox, P. M. (2010). Maximum entropy production in environmental and ecological systems. *Philosophical Transactions of the Royal Society B, 365*, 1297–1302.

Lorente, S., & Bejan, A. (2019). Current trends in constructal law and evolutionary design. *Heat Transfer-Asian Research, 48*, 3574–3589. https://doi.org/10.1002/htj.21556

Nanson, G. C., & Huang, H. Q. (2008). Least action principle, equilibrium states, iterative adjustment and the stability of alluvial channels. *Earth Surface Processes and Landforms, 33*, 923–942.

Nanson, G. C., & Huang, H. Q. (2018). A philosophy of rivers: Equilibrium states, channel evolution, teleomatic change and the least action principle. *Geomorphology, 302*, 3–19.

Ozawa, H., Ohmura, A., Lorenz, R. D., & Pujol, T. (2003). The second law of thermodynamics and the global climate system: A review of the maximum entropy production principle. *Reviews of Geophysics, 41*. https://doi.org/10.1029/2002RG000113

Paik, K., & Kumar, P. (2010). Optimality approaches to describe characteristic fluvial patterns on landscapes. *Philosophical Transactions of the Royal Society B, 365*, 1387–1395.

Patten, B. C. (1995). Network integration of ecological extremal principles: Exergy, energy, power, ascendancy, and indirect effects. *Ecological Modelling, 79*, 75–84.

Pausas, J. G., & Bond, W. J. (2022). Feedbacks in ecology and evolution. *Trends in Ecology and Evolution, 37*, 637–644. https://doi.org/10.1016/j.tree.2022.03.008

Phillips, J. D. (1990). The instability of hydraulic geometry. *Water Resources Research, 26*, 739–744.

Phillips, J. D. (1991). Multiple modes of adjustment in unstable river channel cross- sections. *Journal of Hydrology, 123*, 39–49.

Phillips, J. D. (2011). Emergence and pseudo-equilibrium in geomorphology. *Geomorphology, 132*, 319–326.

Phillips, J. D. (2021). *Landscape evolution. Landforms, ecosystems, and soils*. Elsevier.

Quijano, J., & Lin, H. (2014). Entropy in the critical zone: A comprehensive review. *Entropy, 16*, 3482–3536.

Robertson, D. S. (1991). Feedback theory & Darwinian evolution. *Journal of Theoretical Biology, 152*, 469–484.

Sharma, V., & Annila, A. (2007). Natural process—Natural selection. *Biophysical Chemistry, 127*, 123–128.

Ueltzhöffer, K., Da Costa, L., Cialfi, D., & Friston, K. (2021). A drive towards thermodynamic efficiency for dissipative structures in chemical reaction networks. *Entropy, 23*, 1115. https://doi.org/10.3390/e23091115

Vellend, M. (2016). *The theory of ecological communities*. Princeton University Press, 229 p.

Vermeij, G. J. (1987). *Evolution and escalation*. Princeton University Press.

Vermeij, G. (2011). A historical conspiracy: Competition, opportunity, and the emergence of direction in history. *Cliodynamics: The Journal of Quantitative History and Cultural Evolution, 2*, 187–207.

Wong, M. L., Cleland, C. E., Arend, D., Jr., et al. (2023). On the roles of function and selection in evolving systems. *Proceedings of the National Academy of Sciences (USA), 102*, e2320223120.

Yen, J. D. L., Paganin, D. M., Thomson, J. R., & MacNally, R. (2014). Thermodynamic extremization principles and their relevance to ecology. *Austral Ecology, 39*, 619–632.

3 Forms of Selection in Geophysics

Abstract

This review of abiotic selection covers the role of geophysical phenomena in selection involving biotic/abiotic interactions, and purely or primarily abiotic selection. In evolutionary biology, abiotic selection is the selection pressure exerted on organisms by the geophysical environment. Abiotic effects are important, and often paramount, in ecological filtering, whereby habitat characteristics allow and facilitate some species to establish and prevent or inhibit others. Ecological opportunity occurs when new or modified niches are created, triggering adaptive radiation in biological evolution. Biological evolution can also lead to geological innovation in the form of new features or materials associated with organisms. Ecosystem engineering occurs when organisms modify the environment to affect habitat quality, including modification of abiotic factors, often with evolutionary implications via niche construction. Biogeochemical selection favors faster and more efficient cycling in ecosystem development. The four main forms of purely or mainly abiotic selection are gradient, resistance, network, and thermodynamic selection. *Gradient selection* can be characterized as: the steepest flux gradients usually win, and those pathways are often reinforced by positive feedbacks. *Resistance selection* works by preferential preservation—less resistant features are removed more rapidly, leaving more resistant and stable ones behind. *Network selection* occurs when the formation of specific flow and interaction networks is favored, the main example being branching dendritic flow and circulation networks. Processes that deplete (equalize) energy gradients as rapidly as possible are subject to positive *thermodynamic selection*. Selection associated with the least action principle and maximum entropy production are the primary examples.

J. D. Phillips, *Abiotic Selection in Earth Surface Systems*, Geophysics and Environmental Physics, https://doi.org/10.1007/978-3-031-85862-8_3

3.1 Biotic/Abiotic Interactions

Before presenting some examples of the major types of geophysical selection that are independent of biota, let's examine some common forms of selection involving the interactions of biological and abiotic factors in Earth surface systems (ESS).

3.1.1 Abiotic Selection in Biological Evolution

Evolutionary biologists refer to abiotic selection in terms of non-biological factors that exert selective pressure on organisms. Geophysical and geochemical factors such as temperature, moisture, insolation, substrate stability, chemistry of air, water, soil, and rocks, and physical stresses and disturbances have profound impacts on biota, and the adaptations of organisms to these factors are an important aspect of evolutionary fitness.

Concepts of coevolution in evolutionary biology include not just the coevolution of biota (e.g., between pollinators and flowering plants, or predator-prey "arms races"), but also coevolution of biota and abiotic factors. The evolution of Earth's atmosphere, for example, is reflected in both paleovegetation and in contemporaneous paleosols, reflecting the dependence of atmospheric chemistry on plants (and other life forms) and soils, and coevolution of the atmosphere, biosphere, and pedosphere (Rye & Holland, 1998, 2000; Dergacheva, 2003).

A well-studied example is the coevolution of plants, herbivores, and soils in grassland environments, triggered by factors such as fire, climate change, and atmospheric chemistry, as well as ecological interactions with woody plants (Strömberg, 2011; Retallack, 2013). Other examples involve microbes and intertidal landforms (dating back to the pre-Cambrian; Van der Visjel et al., 2020), and woody vegetation and meandering rivers (Davies & Gibling, 2010, 2013).

"Ancient life and moving fluids" is the title of a paper by Gibson et al. (2021), linking geophysical and computational fluid dynamics with the paleobiology and paleoecology of the Ediacaran biota (635 Ma to the beginning of the Cambrian, ca. 539 Ma). These organisms comprise an assemblage of Neoproterozoic soft-bodied animals representing the first major radiation of macroscopic eukaryotes. Gibson et al. (2021) show how application of fluid dynamics principles advances understanding of Ediacaran paleobiology and discuss a broader perspective of how living things have evolved complex adaptations to navigate and interact with fluid environments.

There is evidence for an even broader role for abiotic selection. The origin of life and biological evolution in the Archean can largely be explained, Michaelian (2011, 2023) argued, in terms of energy dissipation, specifically photon dissipation of solar radiation in the ultraviolet-C wavelengths. He invokes "natural thermodynamic selection" of characteristics of structures and processes, based on increasing the global rate of dissipation of the prevailing solar spectrum.

3.1.2 Ecological Filtering

Ecological filtering, as mentioned earlier, occurs where environmental characteristics—often partly or wholly abiotic—allow or facilitate some species and inhibit or prohibit others. Ecological selection in deserts, for example, filters out organisms that cannot abide water shortages, and provides comparative advantages for those well adapted to dryness. Organism survival depends on minimum amounts of certain resources they must obtain; good health and reproductive success may require higher resource availability. Survival also depends on tolerance limits for stresses such as cold or heat, wetness or aridity, and disturbances (Fig. 3.1). More optimal ranges allow them to do more than merely endure.

Natural selection produces biota with varying degrees of adaptation to environmental conditions. The environment sieves out non- or poorly adapted individuals by subjecting them to unhealthy or lethal stresses and by denying them resources. Cold high-latitude and high-altitude environments, for example, filter out organisms that can't stand freezing conditions. Abiotic factors are commonly, but not exclusively, involved in ecological filtering. Plants not adapted to high levels of hydrogen sulfide, for example, are filtered out by the effects of sulfate-reducing bacteria on soil chemistry. Shading by vegetation in forests, for another example, is an ecological filter for plants with varying degrees of shade tolerance.

Fig. 3.1 Ecological filtering. Perpetually wet conditions in this eastern North Carolina swamp exclude plants not adapted to wetness, and favor plants such as these bald cypress trees (*Taxodium distichum*) which are well adapted to saturated soils and standing water

Ecological filtering exists in all landscapes, though it is most obvious where extreme conditions or limitations require special adaptations. Karst areas, for instance, have soils well suited to many plants. However, high soil pH is unfavorable for plants adapted to the more acidic soils found in other humid settings.

3.1.3 Ecological Opportunity and Adaptive Radiation

Not surprisingly, ecological filtering affects natural selection. Adaptive diversification arising from ecological opportunity—availability or emergence of new environmental conditions that facilitate emergence of new variants—has been a feature of evolutionary research since Darwin. While adaptive radiation (diversification from an ancestor that produces descendants adapted to use new or expanded ecological niches) can occur in other cases, ecological opportunity is usually a precursor (Losos, 2010; Stroud & Losos, 2016).

One example arising directly from geological processes is from the Sahul region of New Guineau and Australia, which includes the Earth's largest and highest tropical island and the oldest and driest continent. Extensive geological uplift in New Guinea from the late Miocene onward acted as a species pump, illustrated by Roycroft et al.'s (2022) studies of geologically recent radiation of rodents of the Muridae family (old world mice and rats) spanning New Guinea, Australia, and oceanic islands. Using phylogenomic data, they found that the orogeny and expansion of New Guinea created ecological opportunities and triggered diversification across a continent. Their results concur with models of increased ecological opportunity across Sahul following major geological uplift ca. 5 Ma. Sustained diversification was driven by over-water colonization from the Pleistocene to present, highlighting the importance of fluid transport of biota and propagules in an evolutionary biogeographic context, as well as in contemporary ecological dispersal.

Freshwater fish provide another example of adaptive diversification associated with ecological opportunity related to geological change, and to changes in hydrodynamics. For example, Costello et al. (2003) highlighted the relative importance of current landscape features and postglacial dispersal in driving genetic variation and population differentiation of *Salvelinus confluentus* (bull trout) in Canada. Adaptive diversification of fish was driven by ecological opportunities associated with habitat conditions created by postglacial lakes in North America (Jarvis et al., 2020). Kern and Langerhans (2018) documented adaptive evolutionary responses in response to river flow changes driven by urbanization. Increased peak flow velocities in urbanized streams resulted in more streamlined bodies of *Rhinichthys obtusus* and *Semotilus atromaculatus* (creek chub) relative to populations in rural streams.

3.1.4 Biological Opportunity and Geological Innovation

The biological energy supplied to Earth surface processes exceeds that of tectonics, gravity-driven surficial processes, and abiotic climate-driven processes (Phillips, 2009). According to Dyke et al. (2011), surface life can also affect interior geological processes such as mantle convection, continental crust uplift, and ocean crust recycling. Power estimates show biota generating 264 TW power, much greater than geological processes (e.g., mantle convection = 12 TW). This is attributable to the biosphere's ability to harvest solar energy from sun. Only a small fraction of biotic energy needs to be used for geomorphic impacts affect surface and interior geological processes (Phillips, 2009; Dyke et al., 2011).

Earth history includes cases where biological evolution—the emergence of new taxa—led directly to novel geological features or phenomena. In the late Paleogene, for example, survivors of extinction events proliferated into a variety of fauna, including corals. Further evolutionary events involving corals occurred during the Miocene and the Plio-Pleistocene. The appearance of these new organisms led to the development of previously-absent landforms—coral reefs. The evolutionary synchrony of the sedimentary stratigraphic record with the biosphere is outlined by Davies et al. (2020), who emphasize the sedimentary record as an archive that records the dynamics of abiotic-biotic interactions, not just a repository for geochemical and fossil records.

Forms created by one of the first forms of life, microbial mats and stromatolites, were the first stage in the eventual development of an entirely new class of geological material, the carbonates. Once limestones and other carbonate rocks were present, new landforms associated with karst appeared.

The rise of photosynthetic bacteria about 2.4 Ma triggered the "great oxygenation event" in the Paleoproterozoic when the atmosphere and shallow ocean saw pronounced rises in oxygen concentrations. These changes triggered new regimes of biogeochemical cycles, solar radiation energy balances, and rock weathering, among other things.

3.1.5 Ecosystem Engineering

Ecosystem engineering occurs when organisms modify their external environment in ways that influence habitat quality or resource availability for themselves and/or other biota. These influences often involve geophysical processes and abiotic entities, and may be positive, negative, or neutral with respect to their effects on the engineer organisms.

At the intersection of geosciences and ecology, studies of ecosystem engineering are most prevalent in the subfields of biogeomorphology and ecohydrology, dating back to well before either of those terms came into wide use in the 1990s. Corenblit et al. (2008) outlined how biological evolution and complex biotic-abiotic feedbacks lead to the emergence of landforms and geomorphic landscape patterns. Other reviews and syntheses of ecosystem engineering in biogeomorphology include Butler (1995), Corenblit et al. (2010, 2011, 2021), and Phillips (2016). Overviews in ecohydrology are provided by Tague and Dugger (2010), Savenije and Hrachowitz (2017), Diamond et al. (2021), and Wilcox et al. (2022).

Ecosystem engineering and associated niche construction play a role in many abiotic-biotic interactions and have done so since the emergence of life. Erwin and Tweedt (2012), for example, assessed the role of positive ecosystem engineering in

Ediacaran-Cambrian diversification of Metazoa. Ecological feedbacks were relatively minor through the Ediacaran, but there was a major increase in the Cambrian, mainly via bioturbation and appearance of engineer organisms.

3.1.6 Biogeochemical Selection

Biogeochemical selection was first explicitly invoked by Lapenis (2002), who traced the seeds of the idea to several early twentieth century Russian scientists. Biogeochemical selection occurs when more rapid (efficient) elemental cycling processes and pathways are preferentially formed and reinforced as ecosystems evolve. Faster cycling can confer advantages to individuals and can also provide benefits at the ecological system level. The more rapidly nutrients and water are used, the more different entities can use them—broadly akin to the advantages of keeping money in circulation in a capitalist economy.

Eagleson (2002) treated biogeochemical selection in greater detail in an ecological context, directly linking biogeochemical to natural selection. A connection between natural and biochemical selection was also drawn by Lekevicius (2006), who noted that while individuals are the focus of the former, ecosystems cannot exist without nutrient cycles. Ecosystems, Lekevicius (2002, 2006) argued, determine the overall direction of evolution. Wilkinson (2003) characterized nutrient cycling as an emergent property of the ecological processes required to support carbon-based life. In the phytotarium concept (Verboom & Pate, 2013), specific vertical layers of vegetation and their associated microbial communities construct niches to maximize access to limiting resources of water and nutrients. Focusing on microbes, Doolittle (2017) asserted that biogeochemical cycles are the unit of selection, combining reproduction with "re-production."

3.2 Gradient Selection

We turn now to selection acting directly on abiotic features and processes, starting with gradient selection. This can be characterized as: the steepest flux gradients are preferentially utilized, and those pathways are often reinforced by positive feedbacks. Chapter 4 is devoted to gradient selection, so this section is focused on providing some illustrative examples as a broad overview of gradient selection in ESS.

3.2.1 Runoff Hydrology

The most visible and intuitively apparent examples of gradient selection are in hydrology, hydrogeology, and fluvial geomorphology. In simple terms, when more rain falls than plants can use and soils can hold, it must be redistributed. Apart from some human engineered water systems, where water flows toward money, and some highly local soil moisture dynamics where capillary forces come into play, this

Fig. 3.2 Gradient selection in action. A mountain stream in southwestern Virginia; ice marking groundwater flow paths in limestone, central Kentucky

redistribution is driven by gravity. Water moves along gradients from higher to lower gravitational potential. Flow is most rapid—efficient—along these steeper gradients, which are selected to convey more water than less steep alternatives. Concentrated flows are more efficient than diffuse flows, and are preferentially selected, explaining why diffuse and sheet flows are rare and spatially and temporally concentrated forms of runoff (Fig. 3.2).

Gradients (paths of least resistance relative to driving force) are also affected by the cross-sectional area and volume of available flow paths, as well as roughness or resistance. Thus, particularly in soil and groundwater flows, opening and partings of various types are chosen by gradient selection—rock joints, fractures, and bedding planes; roots and root channels; micro- and macropores, etc. Gradient selection (including the maintenance and enhancement of flow paths by positive feedbacks) thus explains why preferential flow is ubiquitous (even though flow-through-porous media models like D'arcy's Law may sometimes be a reasonable approximation).

3.2.2 Karst Hydrology

Karst areas are typically characterized by high connectivity of surface and groundwater flow paths, as well as by transitions between surface and subsurface landforms and flow regimes (e.g., Jaillet et al., 2004; Ortega-Becerril et al., 2010; Phillips, 2017; Donnaloia et al., 2019; Gökkakya & Gutiérrez, 2022). The switches

Fig. 3.3 Dry valleys in the central Kentucky karst region. These are common in karst areas as vertical and subvertical conduits capture nearly all runoff. Surface water flows or ponds in these valleys only during very wet periods when underground cavities are filled

are easily explained by gradient selection. If a shaft or swallet opens, for instance, any surface water in the vicinity is going down into it, even if (as is not infrequently the case) it is in the middle of a stream channel. If a channel erodes its way into a doline and provides a more efficient path for at least some of the water collecting in the depression than going down the drain or swallet (these features may become partially or wholly plugged), then the channel will capture the flow (Fig. 3.3).

These examples may seem trivially simple, but they illustrate the ubiquity of abiotic gradient selection in ESS. They are also a good illustration of the local nature of selection, as the steepest or most efficient path at a point may not contribute to the overall efficiency of flow networks (e.g., Jerin & Phillips, 2017).

3.2.3 Hillslope Mass Movements

Slope failures and mass movements depend on the ratio of shear stress to the shear strength of hillslope materials. The latter is affected by cohesion, the angle of internal friction of the hillslope materials, and by effective normal stress, which is influenced by water content. The shear stress is

$$\tau = \gamma h(\cos\theta)(\sin\theta) \tag{3.1}$$

Fig. 3.4 Earthflow in the Raukamura Range, North Island, New Zealand

where γ is the specific gravity and h the thickness of the material above the failure plane. θ is the hillslope angle in degrees. Therefore, for given material properties on a slope, movement depends on the slope gradient (Fig. 3.4). From a slope stability perspective this can be viewed as lower slope gradients being selected for stability and preferential retention. From a mass flux viewpoint, steeper slopes are selected for mass movement.

3.2.4 Atmospheric and Ocean Flows

Wind and air flow are driven by pressure gradients, with their direction relative to the planet affected by the Coriolis effect and (near the ground surface) friction (drag force). Pressure patterns are quite dynamic, but global trends and patterns persist which drive global atmospheric (and ocean) circulation. Air flow patterns cannot be semi-permanently imprinted on the landscape, but some persist, at least partly through positive ocean-atmosphere feedbacks. Global climate teleconnections are one example, with persisting and recurring paths reflecting gradient selection and positive feedback (e.g., Hartmann, 2007; Liu & Alexander, 2007; Fig. 3.5). Similar reasoning applies to the global pattern of persistent ocean currents, where gradients associated with a combination of wind stress on the sea surface and density differences associated with temperature and salinity are selected.

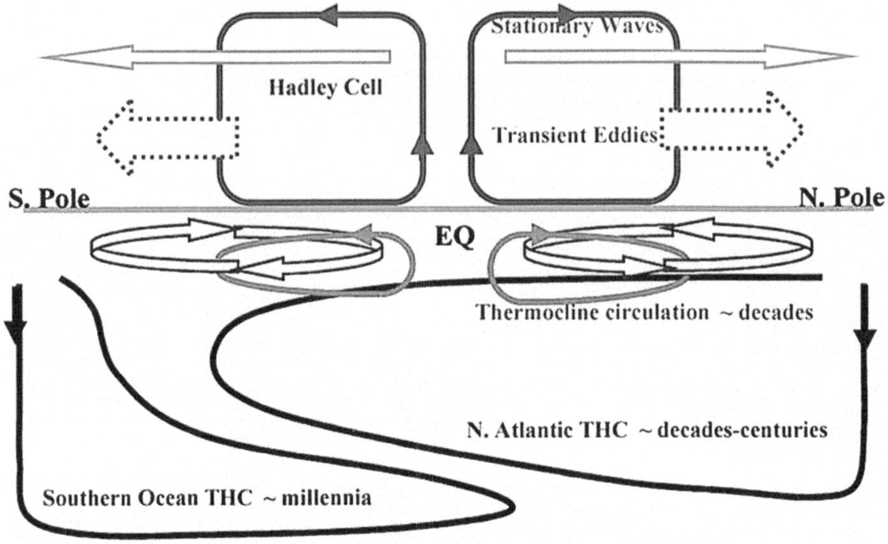

Fig. 3.5 Atmosphere and ocean teleconnections. (Reprinted with permission from Liu and Alexander (2007). © AGU. All rights reserved. Original caption: Schematic for the major branches of climatic teleconnections in the atmosphere and ocean. The atmospheric teleconnections occur at fast timescales, usually shorter than a month (not marked). The oceanic teleconnections occur at a wide range of timescales as marked

3.3 Resistance Selection

Resistance selection works by preferential preservation—less resistant features and materials are removed more readily and rapidly, leaving more resistant and stable ones behind. This is discussed extensively in Chap. 5, so in this section some readily interpreted examples are presented.

Weathering, erosion, denudation, and sediment transport are all controlled by force vs. resistance relationships. For a given regime of driving factors, the more resistant features, materials, and patterns last longer, a form of selection by persistence. Though not traditionally expressed in terms of selection (a notable exception is Scheidegger, 1987, 1991), resistance selection has long been a standard principle in geomorphology. Common examples include preservation of highly resistant rock masses such as igneous batholiths, differential removal and preservation of layered sedimentary rocks, and armoring phenomena, whereby smaller, lighter clasts are preferentially entrained and transported, leaving behind larger and heavier ones (Fig. 3.6). Also ubiquitous is the increasing concentration in soils and regoliths of weathering-resistant residuals—quartz sand grains and iron oxides (typically produced as secondary chemical weathering products) are especially common.

Resistance selection is linked to intrinsic physical and chemical properties of Earth materials, but can also be influenced by factors such as vegetation stabilization of landforms and soils, and various crusting, sealing, and cementing processes in soils and sediments. One example is sand dune fields such as those studied by Hesse (2011). Dunes are normally dynamic, shifting, low-resistance forms.

Fig. 3.6 Resistant selection. (**a**) Bornhardt composed of highly resistant granite—Mt. Ile, Ile-Sede (Errego), Zambezia, Mozambique. Photo by Naomi Oyler, CC-BY-SA 4.0, https://creativecommons.org/licenses/by-sa/4.0/; (**b**) Mushroom rock, White Desert, Egypt—resistant limestone residual showing differential erosion related to resistance differences within the rock. (**c**) Cobble bars, Board Camp Creek, Arkansas. These are a resistant lag deposit with smaller, lighter particles removed. (**d**) Lag gravels, western desert, Egypt, with lighter non-cohesive material removed by wind

However, dating evidence and dune morphology indicate long-term stability of some Australian dunefields. Most of these have experienced only minor superficial modifications since formation up to 1 Ma, notwithstanding large climate changes. "Sticky" dunes (where sand grains are bonded or cemented) may form under a broader range of wind climates than widely thought and have greater resistance to reworking, Hesse (2011) found. Relatively dense vegetation cover compared to more arid deserts may also play a role. Positive feedbacks exist as the relative stability of the dunes allows time for pedogenesis, which results in bonding of sand by pedogenic calcium carbonate, gypsum, silica and translocated clays.

McAuliffe (1994) also identified coevolution of landforms, soil, and vegetation in the Sororan desert of North America. On bajadas, variable resistance leads to spatial variations in erosion and deposition, which in turn create soil contrasts. Spatial variation of soils is associated with variable vegetation cover. Thus, variations in resistance are a strong control of ecological patterning.

3.4 Network Selection

Efficiency selection in ESS often favors the formation of specific flow network types and structures. Chapter 6 will delve into this in more detail, but some examples are provided here.

The task of providing material from numerous sources to a single outlet or sink, or of distributing material from a single source to numerous destinations is facilitated by branching tree-like networks. In many cases these are the most efficient way to achieve these tasks. Thus, developmental trends that move a system toward such networks tend to be favored by selection, and once formed (though their nodes and paths may not be permanently fixed) such networks tend to persist. The efficiency of branching tree networks is well established for information flows (including computer networks and data structures) and biological flow systems, such as circulatory systems in animals and plants. The best-known examples are fluvial channel networks, which in the absence of geological constraints, and given sufficient time, almost invariably form dendritic networks (e.g., Woldenberg, 1969; Rodriguez-Iturbe & Rinaldo, 1997; Dodds, 2010; Fig. 3.7).

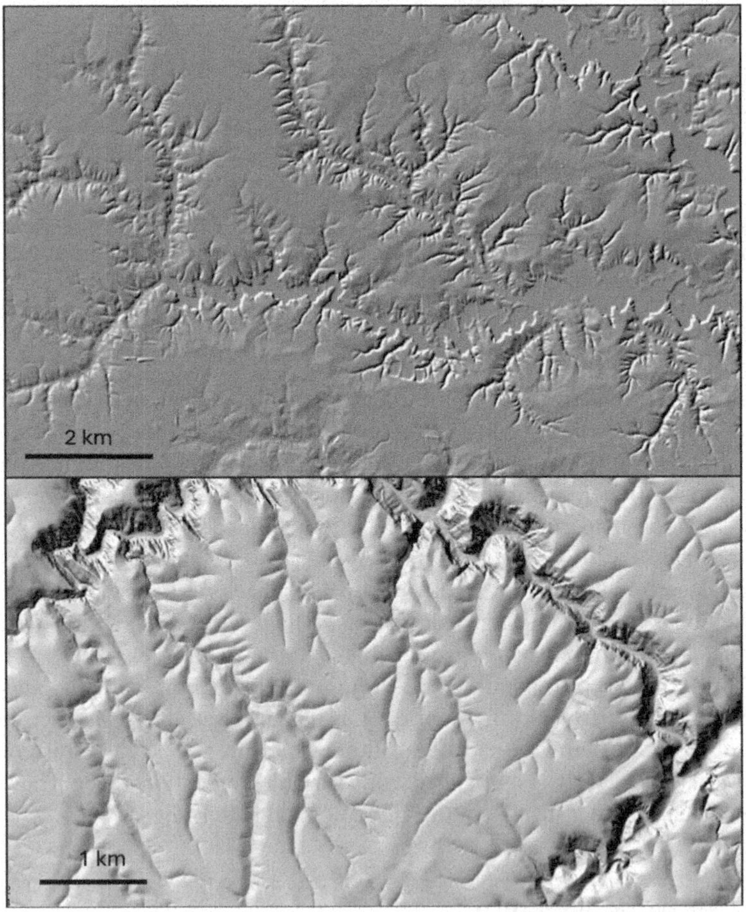

Fig. 3.7 Dendritic stream channel networks as they appear in shaded relief maps in contrasting geological and climate settings. Top: Drainage into Lake Prince on the humid subtropical Virginia coastal plain. Bottom: Portion of the West Clear Creek watershed on the arid Colorado Plateau near Camp Verde, Arizona

Hunt et al. (2023) also demonstrated network selection in the development of groundwater flow paths. Linking network and thermodynamic selection (see below), Ueltzhöffer et al. (2021) identified network selection in chemical energy dissipation. Other examples of network selection, from a constructal law perspective, include snowflake structures and pore networks in porous media (Bejan & Lorente, 2013; Lorente & Bejan, 2019).

3.5 Thermodynamic Selection

Processes or structures that deplete (equalize) energy gradients as rapidly as possible are subject to positive selection. Selection associated with the LAP and maximum entropy production are the primary examples. Links among the laws of thermodynamics, energy dissipation, least action, and maximum entropy have already been mentioned and will be explored in more detail in Chap. 7.

Global atmospheric circulation and poleward heat transport, mentioned in Chap. 1, is perhaps the single most important manifestation of thermodynamic selection. Other examples explored later include climate-ocean interactions such as El Niño-Southern Oscillation, estuarine mixing, fluid turbulence in general and the Navier-Stokes equation in particular, evaporation, and impacts of climate change on river dynamics.

Thermodynamic principles are often vague with respect how selection occurs and how efficiency is defined. In large, complicated systems energy flows can be considered to follow different paths, owing to random variation or complex interactions. Of those paths, the ones associated with more rapid depletion of gradients persist and may intensify. Those that do not generally die out. Thus, varied energetic pathways are selected for or against. Thermodynamic selection therefore reflects the VSR criteria: variations in possible trajectories or evolutionary pathways occur; some are selected; and the ones selected for tend to be retained or to repeatedly recur.

References

Bejan, A., & Lorente, S. (2013). Constructal law of design and evolution: Physics, biology, technology, and society. *Journal of Applied Physics, 113*, 151301.

Butler, D. R. (1995). *Zoogeomorphology: Animals as geomorphic agents*. Cambridge University Press.

Corenblit, D., Gurnell, A. M., Steiger, J., & Tabacchi, E. (2008). Reciprocal adjustments between landforms and living organisms: Extended geomorphic evolutionary insights. *Catena, 73*, 261–273.

Corenblit, D., Steiger, J., & Delmotte, S. (2010). Abiotic, residual and functional components of landforms. *Earth Surface Processes and Landforms, 35*, 1744–1750.

Corenblit, D., Baas, A. C. W., Bornette, G., et al. (2011). Feedbacks between geomorphology and biota controlling earth surface processes and landforms: A review of foundation concepts and current understandings. *Earth-Science Reviews, 106*, 307–331.

Corenblit, D., Corbara, B., & Steiger, J. (2021). Biogeomorphic eco-evolutionary feedback between life and geomorphology: A theoretical framework using fossorial mammals. *The Science of Nature, 108*, 1–16.

Costello, A. B., Down, T. E., Pollard, S. M., et al. (2003). The influence of history and contemporary stream hydrology on the evolution of genetic diversity within species: An examination of microsatellite DNA variation in bull trout, Salvelinus confluentus (Pisces : Salmonidae). *Evolution, 57*, 328–344.

Davies, N. S., & Gibling, M. R. (2010). Cambrian to Devonian evolution of alluvial systems: The sedimentological impact of the earliest land plants. *Earth-Science Reviews, 98*, 171–200.

Davies, N. S., & Gibling, M. R. (2013). The sedimentary record of Carboniferous rivers: Continuing influence of land plant evolution on alluvial processes and Palaeozoic ecosystems. *Earth-Science Reviews, 120*, 40–79.

Davies, N. S., Shillito, A. P., Slater, B. J., et al. (2020). Evolutionary synchrony of Earth's biosphere and sedimentary-stratigraphic record. *Earth-Science Reviews, 201*, 102979.

Dergacheva, M. (2003). Pedohumic method in paleoenvironmental reconstructions: An example from middle Siberia. *Quaternary International, 106*, 73–78.

Diamond, J. S., Epstein, J. M., Cohen, M. J., et al. (2021). A little relief: Ecological functions and autogenesis of wetland microtopography. *Wiley Interdisciplinary Reviews—Water, 8*, e1493. https://doi.org/10.1002/wat2.1493

Dodds, P. S. (2010). Optimal form of branching supply and collection networks. *Physical Review Letters, 104*, 048702.

Donnaloia, M., Giachetta, E., Capolongo, D., & Pennetta, L. (2019). Evolution of fluviokarst canyons in response to the Quaternary tectonic uplift of the Apulia Carbonate Platform (Southern Italy): Insights from morphometric analysis of drainage basins. *Geomorphology, 336*, 18–30. https://doi.org/10.1016/j.geomorph.2019.03.008

Doolittle, W. F. (2017). Darwinizing Gaia. *Journal of Theoretical Biology, 434*, 11–19.

Dyke, J. G., Gans, F., & Kleidon, A. (2011). Towards understanding how surface life can affect interior geological processes: A non-equilibrium thermodynamics approach. *Earth System Dynamics, 2*, 139–160.

Eagleson, P. S. (2002). *Ecohydrology: Darwinian expression of vegetation form & function*. Cambridge University Press.

Erwin, D. H., & Tweedt, S. (2012). Ecological drivers of the Edicarian-Cambrian diversification of Metazoa. *Evolutionay Ecology, 26*, 417–433.

Gibson, B. M., Furbish, D. J., Rahman, I. A., et al. (2021). Ancient life and moving fluids. *Biological Reviews, 96*, 129–152.

Gökkakya, E., & Gutiérrez, F. (2022). Poljes in the Sivas gypsum karst, Turkey. *Geomorphology, 417*, 108451. https://doi.org/10.1016/j.geomorph.2022.108451

Hartmann, D. L. (2007). The atmospheric general circulation and its variability. *Journal of the Royal Meteorological Society of Japan, 85B*, 123–143.

Hesse, P. (2011). Sticky dunes in a wet desert: Formation, stabilisation and modification of the Australian desert dunefields. *Geomorphology, 134*, 309–325. https://doi.org/10.1016/j.geomorph.2011.07.008

Hunt, A. G., Sahimi, M., & Ghanbarian, B. (2023). Predicting streamflow elasticity based on percolation theory and ecological optimality. *AGU Advances, 4*, e2022AV000867. https://doi.org/10.1029/2022AV000867

Jaillet, S., Pons-Branchu, E., Brulhet, J., & Hamelin, B. (2004). Karstification as geomorphological evidence of river incision: The karst of Cousance and the Marne valley (eastern Paris Basin). *Terra Nova, 16*, 167–172.

Jarvis, W. M. C., Peiman, K. R., Cooke, S. J., & Robinson, B. W. (2020). Low connectivity between sympatric populations of sunfish ecotypes suggests ecological opportunity contributes to diversification. *Evolutionary Ecology, 34*, 391–410.

Jerin, T., & Phillips, J. D. (2017). Local efficiency in fluvial systems: Lessons from Icicle Bend. *Geomorphology, 282*, 119–130.

Kern, E. M. A., & Langerhans, R. B. (2018). Urbanization drives contemporary evolution in stream fish. *Global Change Biology, 24*, 3791e3803.

Lapenis, A. G. (2002). Directed evolution of the biosphere: Biogeochemical selection or Gaia? *Professional Geographer, 54*, 379–391. https://doi.org/10.1111/0033-0124.00337

Lekevicius, E. (2002). *The origin of ecosystems by means of natural selection*. Lithuanian Academy of Sciences, Institute of Ecology.
Lekevicius, E. (2006). The Russian Paradigm in ecology and evolutionary biology: Pro et contra. *Acta Zoologica Lituanica, 16*, 3–19.
Liu, Z., & Alexander, M. (2007). Atmospheric bridge, oceanic tunnel, and global climatic teleconnections. *Reviews of Geophysics, 45*, RG2005. https://doi.org/10.1029/2005RG000172
Lorente, S., & Bejan, A. (2019). Current trends in constructal law and evolutionary design. *Heat Transfer–Asian Research, 48*, 3574–3589. https://doi.org/10.1002/htj.21556
Losos, J. B. (2010). Adaptive radiation, ecological opportunity, and evolutionary determinism. *American Naturalist, 175*, 623–639. https://doi.org/10.1086/652433
McAuliffe, J. R. (1994). Landscape evolution, soil formation, and ecological patterns and processes in Sonoran desert bajadas. *Ecological Monographs, 64*, 11148.
Michaelian, K. (2011). Thermodynamic dissipation theory for the origin of life. *Earth System Dynamics, 2*, 37–51. https://doi.org/10.5194/esd-2-37-2011
Michaelian, K. (2023). The non-equilibrium thermodynamics of natural selection: From molecules to the biosphere. *Entropy, 25*, 1059.
Ortega-Becerril, J. A., Heydt, G. G., & Duran, J. J. (2010). The role of sculpted forms and endokarstic active conduits in the development of fluviokarstic canyons. The Rio Puron cave conduit (Spain). In B. Andreo (Ed.), *Advances in research in karst media* (pp. 387–392). Springer.
Phillips, J. D. (2009). Biological energy in landscape evolution. *American Journal of Science, 309*, 271–290.
Phillips, J. D. (2016). Biogeomorphology and contingent ecosystem engineering in karst landscapes. *Progress in Physical Geography, 40*, 503–526. https://doi.org/10.1177/0309133315624641
Phillips, J. D. (2017). Landform transitions in a fluviokarst landscape. *Zeitchschrift für Geomorphologie, 61*(2), 109–122.
Retallack, G. J. (2013). Global cooling by grassland soils of the geological past and near future. *Annual Review of Earth and Planetary Sciences, 41*, 69–86.
Rodriguez-Iturbe, I., & Rinaldo, A. (1997). *Fractal River Basins. Chance and self-organization*. Cambridge University Press, New York.
Roycroft, E., Fabre, P. H., MacDonald, A. J., et al. (2022). New Guineau uplift opens ecological opportunity across a continent. *Current Biology, 32*, 4215. https://doi.org/10.1016/j.cub.2022.08.021
Rye, R., & Holland, H. D. (1998). Paleosols and the evolution of atmospheric oxygen: A critical review. *American Journal of Science, 298*, 621–672.
Rye, R., & Holland, H. D. (2000). Life associated with a 2.76 Ga ephemeral pond? Evidence from Mount Roe #2 paleosol. *Geology, 28*, 483–486.
Savenije, H. H. G., & Hrachowitz, M. (2017). HESS opinions: Catchments as meta-organisms—A new blueprint for hydrological modeling. *Hydrology and Earth System Sciences, 21*, 1107–1116.
Scheidegger, A. E. (1987). The fundamental principles of landscape evolution. *Catena Supplement, 10*, 199–210.
Scheidegger, A. E. (1991). *Theoretical geomorphology* (3rd ed.). Springer.
Strömberg, C. A. E. (2011). Evolution of grasslands and grassland ecosystems. *Annual Review of Earth and Planetary Sciences, 39*, 517–544.
Stroud, J. T., & Losos, J. B. (2016). Ecological opportunity and adaptive radiation. *Annual Review of Ecology, Evolution, and Systematics, 47*, 507–532.
Tague, C., & Dugger, A. L. (2010). Ecohydrology and climate change in the mountains of the western USA – A review of research and opportunities. *Geography Compass, 4*, 1648–1663. https://doi.org/10.1111/j.1749-8198.2010.00400.x
Ueltzhöffer, K., Da Costa, L., Cialfi, D., & Friston, K. (2021). A drive towards thermodynamic efficiency for dissipative structures in chemical reaction networks. *Entropy, 23*, 1115. https://doi.org/10.3390/e23091115

van der Visjel, R. C., van Belzen, J., & Bouma, T. J. (2020). Estuarine biofilm patterns: Modern analogues for Precambrian self-organization. *Earth Surface Processes and Landforms, 45*, 1141–1154.

Verboom, W. H., & Pate, J. S. (2013). Exploring the biological dimension to pedogenesis with emphasis on the ecosystems, soils, and landscapes of southwestern Australia. *Geoderma, 211*(2), 154–183.

Wilcox, B. P., Basant, S., Olariu, H., & Leite, P. A. M. (2022). Ecohydrological connectivity: A unifying framework for understanding how woody plant encroachment alters the water cycle in drylands. *Frontiers in Environmental Science, 10*, 934535. https://doi.org/10.3389/fenvs.2022.934535

Wilkinson, D. M. (2003). The fundamental processes in ecology: A thought experiment on extra-terrestrial biospheres. *Biological Reviews, 78*, 171–179.

Woldenberg, M. J. (1969). Spatial order in fluvial systems: Horton's laws derived from mixed hexagonal hierarchies of drainage basin areas. *Geological Society of America Bulletin, 80*, 97–112.

4 Gradient Selection

Abstract

Gradient selection refers to the flows of mass and energy occurring along the steepest gradients of potentials or concentration. This is most intuitive for gravity-driven flows, but it also applies to wind and air flows due to gradients of atmospheric pressure, heat flux via conduction from hotter to colder, and chemical diffusion from higher to lower concentrations. Examples of gradient selection described here include wind and air flow patterns in the atmosphere, hillslope mass movements, topographic steering of wind in sand dunes, and karst dissolution dynamics. Hydraulic selection is a form of gradient selection applicable to fluvial systems and is outlined here in the context of the "job of the river" to move excess water, with geomorphic work of the river a byproduct of its hydrologic job. Preferential gradient flow paths generally persist in situations where flows can form and modify their pathways due to positive feedback (e.g., preferential flow paths in soil, groundwater, and surface hydrology), and tend to recur or persist in other situations (e.g., ocean currents and global wind patterns). Gradient selection is local, such that accessing the steepest gradient in the immediate vicinity may or may not enhance the flux efficiency of the larger system. Examples include stream capture and ice stream piracy.

4.1 Selection of the Steepest

Mass and energy fluxes occur along the steepest gradients of potentials or concentration. This is most intuitive for gravity-driven flows of water, sediment, and hillslope materials. But it also applies to wind and air flows due to gradients of atmospheric pressure, heat flux via conduction from hotter to colder, and chemical diffusion from higher to lower concentrations. The principle of gradient selection in geomorphology is simply that geomorphic features associated with these gradients

J. D. Phillips, *Abiotic Selection in Earth Surface Systems*, Geophysics and Environmental Physics, https://doi.org/10.1007/978-3-031-85862-8_4

persist and grow (Phillips, 2011). Hydraulic selection principles operate, for instance, as water is redistributed on the surface. The most efficient paths persist and prevail; less efficient options dry up. As channels are scoured due to initially chance concentrations of shear stress relative to local resistance, hydraulic selection favors channels over diffuse flow, and favors the steepest, largest, and smoothest channels over smaller, flatter, and rougher ones. Twidale (2004, p. 170) characterized this as a process of natural selection.

The principle also applies to other aspects of geophysics and Earth sciences. Gradient selection is embedded in the constructal law (Bejan, 2007) as applied to phenomena such as turbulent shear layers, jets, and plumes, vortex shedding, and Rayleigh-Bénard convection. An important distinction is between gradient-driven flows that inherently deplete those gradients and those that do not necessarily do so. Heat flux, and air flow along pressure gradients deplete those gradients by equalizing the differences as heat flows from warmer to colder and air moves from high to low pressure. Others, such as gravity-driven water and mass flows, do not necessarily deplete the topographic or gravitational gradients. Work performed because of those gradient fluxes, such as erosion from higher and deposition at lower elevations can deplete the gradients, but the fluid flows absent their byproducts of geomorphic work do not. The issue of gradient depletion is revisited in Chap. 7.

Channelized flows and surface runoff provide the most obvious examples of gradient selection, but there are numerous others. Gradient selection is reflected in the prevalence of preferential hydrological flow paths at a range of scales from soil pores to continental, and the tendency for these to either be controlled by, or to evolve into, morphological features such as rills, gullies, stream channels, conduits, and macropores in rock, sediments, soils, and landscapes, (Lin, 2010; Kleidon et al., 2013). Water fluxes follow paths of least resistance such as macropores, fractures, and zones of higher hydraulic conductivity, but the flows are driven by gradients (in this case gravity). The flow powers weathering, dissolution, erosion, and sediment transport within these features, which enhances these pathways at the expense of other nearby zones. Even in apparently homogeneous materials without evident weaknesses or potential flow routes, unstable growth of minor initial variations in moisture content causes these to grow by positive feedback, largely via higher conductivities in locally saturated zones. Preferential water flow along or initiated by pre-existing physical features is often reinforced by preferential root penetration, and other biological activity along the same routes.

4.1.1 Gradient Selection in the Atmosphere

Gradient-driven fluxes will select the strongest (steepest) gradient. In the atmosphere, pressure gradients drive wind and air flow.

The hydrostatic equation gives the change in pressure p with height z based on air density ρ:

$$\frac{\Delta p}{\Delta z} = -\rho g \tag{4.1}$$

The pressure gradient between two locations is $PG = \Delta p/\Delta d$, where Δd is the distance between them. The magnitude of the pressure gradient force per meter of distance at a given height is

$$\frac{F_{PG}}{m} = \left|\frac{\Delta p}{\rho \Delta d}\right|. \tag{4.2}$$

At a given pressure (i.e., on a constant pressure map) the force is

$$\frac{F_{PG}}{m} = \left|-g\frac{\Delta Z}{\Delta d}\right| \tag{4.3}$$

where Z is the difference in height contours.

F_{PG} is often calculated for both north-south (meridional, y) and east-west (zonal, x) components:

$$\frac{F_{xPG}}{m} = -\frac{1}{\rho}\frac{\Delta p}{\Delta x} \tag{4.4a}$$

$$\frac{F_{yPG}}{m} = -\frac{1}{\rho}\frac{\Delta p}{\Delta y} \tag{4.4b}$$

$$\frac{F_{xPG}}{m} = -g\frac{\Delta Z}{\Delta x} \tag{4.4c}$$

$$\frac{F_{yPG}}{m} = -g\frac{\Delta Z}{\Delta y} \tag{4.4d}$$

Wind is also influenced by centrifugal force (F_{CN}) for a curving air parcel, varying as a direct function of wind velocity M squared, and as an inverse function of the radius of curvature R, and turbulent drag (see below), a function of velocity and the intensity of turbulence.

$$\frac{F_{xCN}}{m} = s\frac{vM}{R} \tag{4.5a}$$

$$\frac{F_{yCN}}{m} = -s\frac{uM}{R} \tag{4.5b}$$

Where u, v are the meridional and zonal components of the wind, and s is 1 for winds circling high pressure in the southern or low pressure in the northern hemisphere, and $s = -1$ for winds circling low pressure in the southern or high pressure in the northern hemisphere.

The Coriolis effect acts to the left of the wind in the southern and the right of the wind in the northern hemisphere. It is proportional to wind speed and affects only the direction of the wind, with larger effects at higher latitudes and none at the equator.

$$\frac{F_{xCF}}{m} = fv \tag{4.6a}$$

$$\frac{F_{yCF}}{m} = -fu \tag{4.6b}$$

The Coriolis parameter *f* is

$$f = 2\Omega \sin \tag{4.7}$$

where Ω is Earth's rotation rate and ø is latitude (positive for north, negative for south).
In the atmospheric boundary layer turbulent drag or friction force is also present:

$$\frac{F_{xTD}}{m} = -C_d u \tag{4.8a}$$

$$\frac{\mathrm{F}_{yTD}}{m} = -C_d v \tag{4.8b}$$

C_d is a drag coefficient that varies according to the intensity of turbulence.

Wind can move air at a different velocity past a fixed location that causes wind at the fixed point to change over time. This is called advective force (though it is not a true force) and is a function of the velocities in the meridional, zonal, and vertical directions.

$$\frac{F_{xAD}}{m} = -u\frac{\Delta u}{\Delta x} - v\frac{\Delta u}{\Delta y} - w\frac{\Delta u}{\Delta z} \tag{4.9a}$$

$$\frac{F_{yAD}}{m} = -u\frac{\Delta v}{\Delta x} - v\frac{\Delta v}{\Delta y} - w\frac{\Delta v}{\Delta z} \tag{4.9b}$$

The last term on the right, representing the vertical component, is generally small and can often be neglected.

All the factors above figure into the equations of motion for horizontal air movements in the atmospheric boundary layer, based on Newton's second law. But only the pressure gradient force can cause air that is initially at rest to begin moving; the others come into play once motion is initiated. Considering all the forces and effects the horizontal equations of motion are:

$$\frac{\Delta u}{\Delta t} = -u\frac{\Delta u}{\Delta x} - v\frac{\Delta u}{\Delta y} - w\frac{\Delta u}{\Delta z} - \frac{1}{\rho}\frac{\Delta p}{\Delta x} + fv - C_D u \tag{4.10a}$$

$$\frac{\Delta v}{\Delta t} = -u\frac{\Delta v}{\Delta x} - v\frac{\Delta v}{\Delta y} - w\frac{\Delta v}{\Delta z} - \frac{1}{\rho}\frac{\Delta p}{\Delta y} + fu - C_D v \tag{4.10b}$$

Wind velocities and directions can be quite ephemeral as regional and local pressure gradients shift. However, at the global scale certain pressure patterns related to the variations in solar radiation input persist, creating zones of prevailing winds. This produces general patterns that are familiar in introductory physical geography and meteorology textbooks (Fig. 4.1).

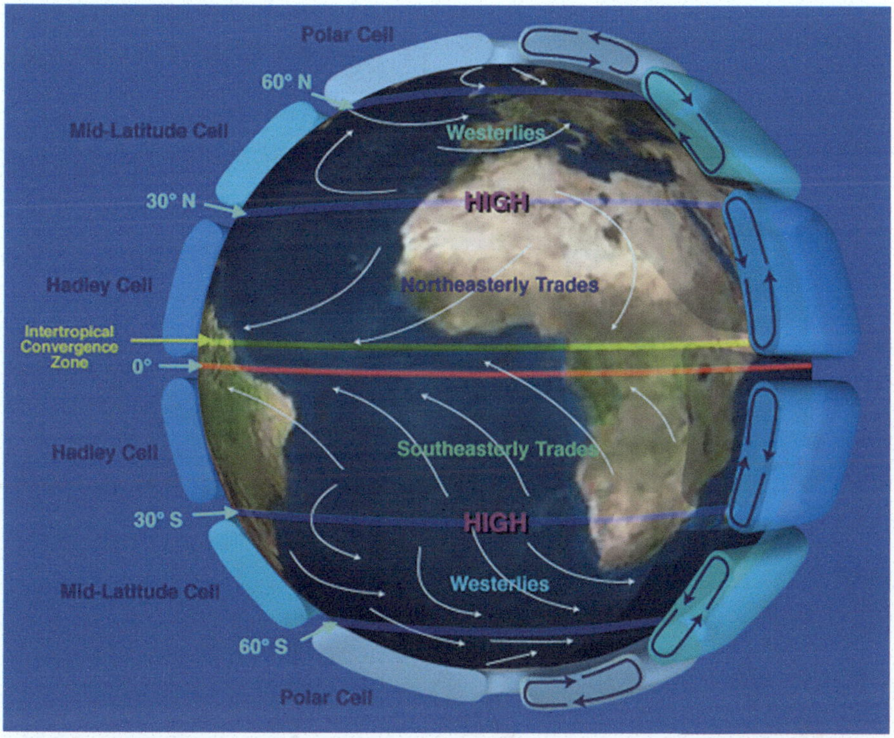

Fig. 4.1 A simplified representation of Earth's general atmospheric circulation patterns. (Source: NASA (public domain))

Interactions among the pressure gradient, centrifugal, and drag forces and the Coriolis effect can be depicted as shown below (Fig. 4.2). The pressure gradient force has a positive effect on centrifugal force through its effects on wind speeds. Likewise, higher velocities increase the Coriolis "force." Other factors being equal, drag force is higher with higher wind speeds. Coriolis effects may influence centrifugal force via curved wind paths, but these vary. Advection is left out because its effects are also situationally contingent. F_{TD} has negative feedbacks on F_{PG} and F_{CN} via velocity, which can be increased by more intense turbulence and greater drag coefficients.

If not for the negative feedback of the drag force, higher (lower) wind velocities associated with increasing (decreasing) pressure gradients would drive the system to dynamical instability. Selection favors stability, and therefore maximization of turbulence so that drag exerts negative feedback to stabilize the wind system. We will revisit this phenomenon in Chap. 7.

Preferred flow paths in the atmosphere are therefore linked to persistent or recurring patterns of pressure that manifest at regional and local as well as global scales—as the pressure gradient force depletes those gradients, recurrent or persistent flow requires that they be rejuvenated. Winds can also be blocked or steered by

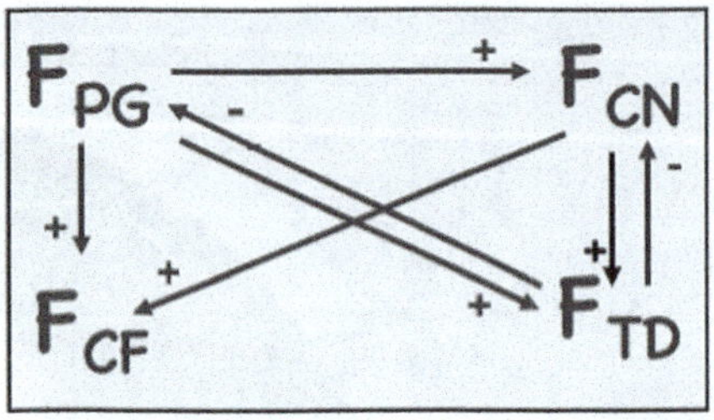

Fig. 4.2 Positive and negative feedbacks among the pressure gradient, centrifugal, Coriolis, and turbulent drag forces in the atmospheric boundary layer

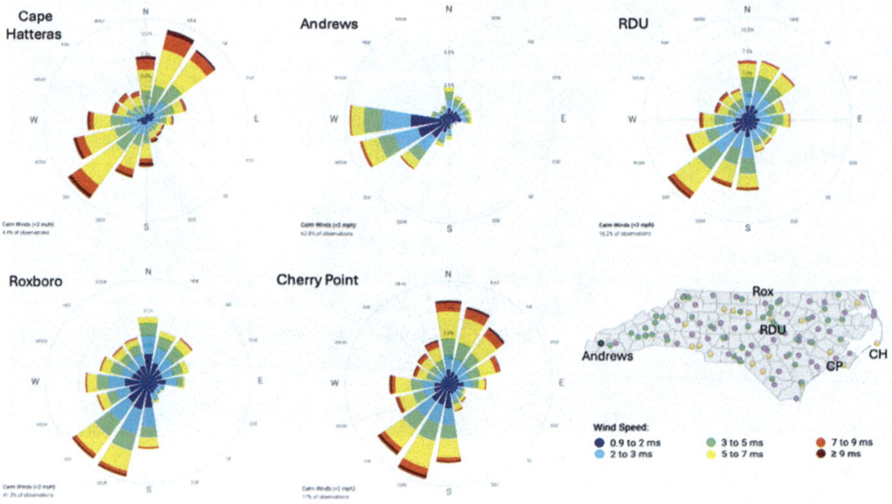

Fig. 4.3 Wind roses for five meteorological stations in North Carolina, produced using a tool developed by the climate center at North Carolina State University (https://airquality.climate.ncsu.edu/wind/). The Cape Hatteras station shows clear trends toward domination by northeasterly and southwesterly winds associated with common synoptic pressure patterns in the region, reflected to a slightly lesser degree at the Marine Corps Air Station Cherry Point site. The Raleigh-Durham airport (RDU) and Roxboro sites have more varied directionality and generally weaker wind velocities, though the regional pattern is evident. At Andrews, in the mountains, topography greatly affects winds

topography to produce persistent patterns of wind direction. Figure 4.3 shows some examples from North Carolina. However, while wind patterns are consistent with gradient selection, recognizing them as such does not add greatly to our understanding. This is because, unlike preferential flows of water or ice, atmospheric flows typically have no mechanism for structural or morphological reinforcement of preferential flow paths—they cannot create or enlarge a gradient flux path, which is usually externally imposed by atmospheric pressure patterns and sometimes topography. An exception is in sand dune environments, where aeolian erosion and deposition has feedbacks on airflow that reinforce or modify pressure gradient flux pathways. One example, that of localized deflation hollows or dune blowouts, is discussed in one of the case studies later in the chapter.

4.1.2 Hillslope Mass Movements

Mass movements on hillslopes occur when and where critical thresholds of shear strength vs. shear stress are exceeded, typically represented using the factor of safety (*F*) equation:

$$F = \frac{c + \sigma' tan\phi}{\gamma h(cos\theta)(sin\theta)} \tag{4.11}$$

The numerator is shear strength given by the Mohr-Coulomb equation as a function of cohesion (c), effective normal stress (σ'), and the angle of internal friction (ϕ). The denominator is shear stress, which depends on specific weight of the mass (γ), thickness of the slope material above the potential failure plane (h), and the slope angle (θ) relative to the horizontal. Force, strength and stress parameters have units of Nm^{-2}.

For a particle or clast on a slope,

$$F = \frac{mg\,cos\theta}{mg\,sin\theta} \tag{4.12}$$

where m is particle mass and g is the gravity constant.

For a given particle or layer of hillslope regolith, movement will only occur where gradients result in $F < 1$. One possibility is that those portions steeper than the threshold will experience failures until the angle of repose (where $F = 1$) is reached (Fig. 4.4). Positive feedback may come into play if the post-failure slopes remain at or near the angle of repose, because small increases in steepness or reductions in shear strength may initiate new failures.

Fig. 4.4 Sand dune at Shackleford Banks, North Carolina. Multiple local sand avalanches have occurred where gradients exceeded the force-resistance threshold. In the upper part of the dune reinforcement by roots increases shear strength so that failures are less common. (Published with permission of Morgan O. Phillips and Nathan S. Phillips, © 2021. All rights reserved)

Fig. 4.5 Earthflow site in the Sumava Mountains, Czech Republic. The transported material collects moisture delivered upslope from the flow path. The frequent saturation results in chronic creep and occasional flow reactivation

Some sites experience repeated landslides or other slope failures (Fig. 4.5). In some cases, this arises from aspects of the environmental setting extrinsic to the slope itself, such as seismic activity or a climate subject to intense rainfalls or storm events. However, self-reinforcing positive feedbacks can also occur. Samia et al. (2017) examined 17 landslides in the Umbria region of Italy and found that overlap in the events demonstrated clear path dependency and legacy effects—i.e., landslides begat landslides. The positive feedback is temporary, however (about 10 years in their study), and diminishes over time.

Mass movement locations are therefore strongly controlled by gradient selection but feedbacks to subsequent failures may be negative or positive.

4.1.3 Topographic Steering in Sand Dunes

Topography can have important effects on wind and air flow at the surface boundary layer. Selection can occur when wind interacts with a deformable boundary, which most commonly occurs in settings such as arid and coastal sand dunes and is particularly important in dune blowouts (also called deflation hollows; Fig. 4.6). Blowouts are localized erosional features that are broadly classified into elongated trough or wider saucer blowouts. In addition to sandy desert and coastal areas, blowouts are the most regionally pervasive active landforms on the northern Great Plains of North America (Hugenholtz & Wolfe, 2006).

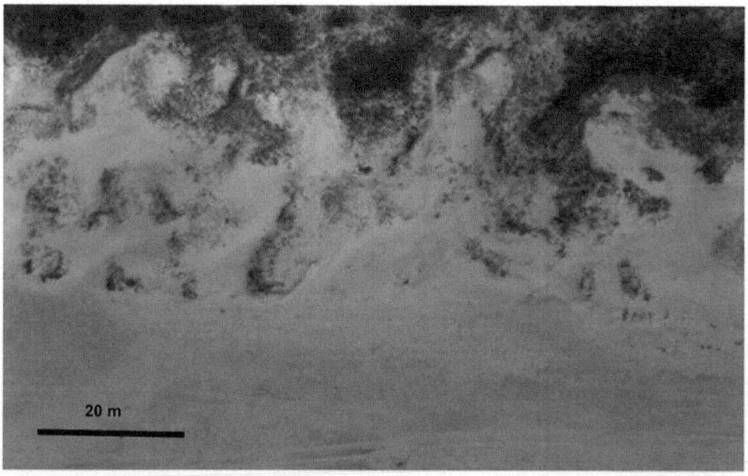

Fig. 4.6 A complex of dune blowouts on Ocracoke Island, North Carolina. Atlantic Ocean is toward the bottom of the image. (Image acquired 15 April 2017 by U.S. National Oceanic and Atmospheric Administration (https://coast.noaa.gov/htdata/NGS_oblique/NGSOblique_2017/2017obP10553112.tif))

Feedbacks among wind flow, dune morphology, sediment transport and deposition, and vegetation were rapid and ongoing in New Jersey coastal dune blowouts (Gares & Nordstrom, 1995), with an entire developmental cycle from initiation to closure occurring in 20 years. The model of Gares and Nordstrom has four phases (Fig. 4.7). First, an erosion notch is initiated in a foredune by offshore winds that are topographically accelerated at the dune crest. The notches develop into incipient blowouts due to deflation in the throat in stage 2, followed by stage 3, where sand is deflated from the throat to a back beach depositional apron, and the blowout is widened by erosion on the side walls. Deflational sediment export ceases and widening slows in stage 4 as the lateral walls develop gentler slopes. This decelerates wind velocities in the blowout, resulting in sand deposition and eventual closure. Hugenholtz and Wolfe (2006) found that saucer-shaped deflation hollows in the northern Great Plains evolved progressively, with stages analogous to stages 1–3, but then often developed into parabolic dunes.

Selection plays a role in at least two ways. First, initial notch formation may exploit local points of low resistance (resistance selection); often where vegetation has been damaged or destroyed. Gradient selection may also play a role here, as minor topographic features that funnel wind may initiate the notches. Selection also occurs in topographic steering of flow and wind/sediment/morphology interactions within a blowout.

Wind approaching a trough blowout parallel to the throat orientation results in jets occurring in the bottom of the feature (deflation basin) and along the walls (Hesp, 1996). Jets also form over the depositional lobe crest, with downwind flow separation on the leeward side of the lobe. Over the crests of the erosional walls, flow separation occurs and corkscrew vortices form. Maximum erosion and

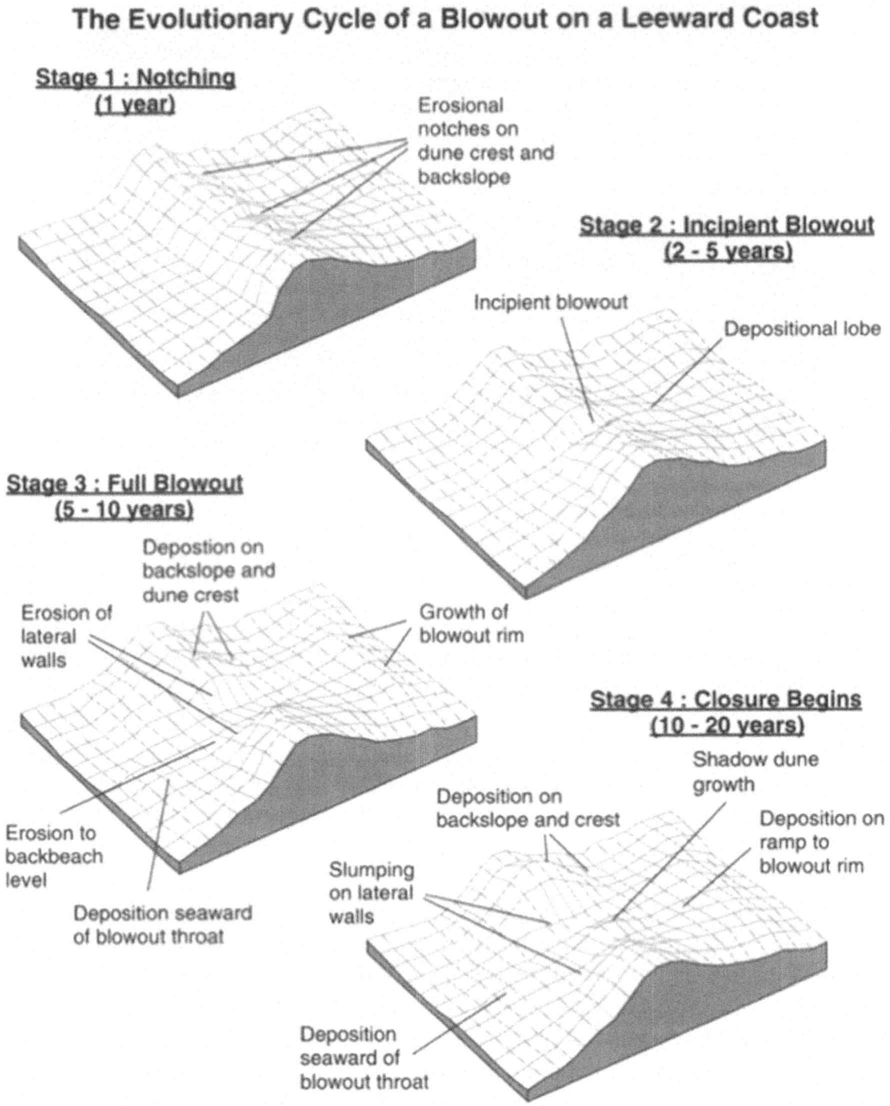

Fig. 4.7 Evolutionary stages of a blowout on a leeward coast. (Reprinted with permission from Gares and Nordstrom (1995).)

transport occur up the deflation basin and onto the depositional lobe. Based on monitoring of a blowout in coastal New South Wales (Australia), Hesp (1996) maintained that trough blowout morphologies are explained as a function of these flow patterns, though there are obvious form-process feedbacks.

Interestingly, winds do not need to be from any specific direction. Blowouts are influenced by both onshore and offshore winds. The general phenomenon of topographic steering of wind into and through the blowout was demonstrated by Hesp (1996), Hesp and Hyde (1996), and Hesp and Pringle (2001). The latter conducted detailed measurements in a trough blowout in New Zealand, roughly 40 m long, 20 m wide, and 7 to 8 m deep. Nearly any wind approaching the feature from seaward, alongshore, or slightly offshore directions in an arc of ≥200 degrees to the seaward entrance orientation was steered into the trough. This occurred due to four factors. The relatively deep topography created intense localized low pressure, sucking wind into the blowout. Second, flow entering the feature was constrained and steered by the blowout walls. Third, local flow separation occurred during alongshore and offshore winds along the blowout. Fourth, flows of low to moderate obliquity are deflected or re-directed seaward, over the foredune.

Saucer blowouts also have strong topographic steering of wind and flow-form interactions, but angle of wind approach results in more variation, and flow-morphology interactions between saucer blowouts may differ more than is the case for trough features (Hugenholtz & Wolfe, 2006, 2009). Mode switches may occur as blowouts grow, with positive feedback dominating smaller saucers, reinforcing deflations from the blowout. At a threshold size, however, negative feedback may inhibit deflation in larger blowouts.

Field measurements integrated with computational fluid dynamics (CFD) were utilized by Jackson et al. (2011) to analyze wind flow/morphology interactions during offshore winds on coastal dunes in Northern Ireland. They found that offshore and obliquely offshore winds result in flow reversal and onshore directed winds at distances of up to 20 m from the dune toe. In a study in the Canary Islands, Smith et al. (2021) found meso-scale topographic steering at two scales—regional airflow bifurcation at the island scale, and local-scale topographic modification at the dune-field scale. Specifically, CFD simulations of northerly winds identified two characteristic meso-scale airflow perturbations: Island-scale bifurcation of airflow around the island, steered parallel to the lateral coastlines before converging at the lee of the island, and smaller scale (i.e., ~2 km) turbulent canyon breezes. Blowout development on artificial dunes differs somewhat from that on natural dunes, but the topographic steering and morphology-wind feedbacks are also critical on constructed dunes (Zhang et al., 2022).

Hugenholtz and Wolfe (2009) examined the details of airflow during a chinook wind event where wind directions rotated, allowing a more detailed assessment of the role of approach angle. They found that wind blowing over the windward rim of the saucer blowout expands and decelerates, triggering flow separation and a small re-circulation zone on sheltered lee slopes. Air flow re-attaches to the blowout surface near the deflation basin and accelerates up to a small opening on the downwind rim. Using a fractional speed-up ratio (δ_s),

$$\delta_s = \frac{(u_z - U_z)}{U_z} \tag{4.13}$$

where u_z is wind velocity at height z on the leeward rim, and U_z the velocity at height z on the reference anemometer upwind, they found acceleration of up to 50%. Smyth et al. (2013) measured and modelled airflow patterns at a range of windspeeds in a bowl-shaped blowout, also finding an increase in wind speed and turbulence intensity relative to incident winds.

Aeolian researchers generally, at least in recent decades, integrate dense instrumental measurements with CFD models. In published studies there generally exists good agreement between models and anemometer data. Various CFD models, both Lagrangian and Eulerian, are employed, with varying assumptions, solution methods, and parameterization schemes. All, however, are based on a logarithmic vertical wind velocity profile and the law of the wall:

$$U(z) = \frac{u_*}{K} \ln\left(\frac{z + z_0}{z_0}\right) \tag{4.14}$$

$U(z)$ is wind speed at height z, u_* is shear velocity, z_0 is aerodynamic surface roughness length, and K is the von Karman constant (≈ 0.4). The “wall” is the outer boundary of the fluid region. The law states that the mean velocity of a turbulent flow is proportional to the (log of the) distance to the wall. Shear velocity is a function of wall shear stress (τ_w) and fluid density (ρ):

$$u_* = \sqrt{\frac{\tau_w}{\rho}} \tag{4.15}$$

Turbulent kinetic energy at the entrance (k) and specific turbulent dissipation rate (ε) are calculated using

$$k = \frac{u_*^2}{\sqrt{C_\mu}} \tag{4.16}$$

$$\varepsilon = \frac{u_*^3}{K(z + z_0)} \tag{4.17}$$

C_μ is a constant. Then, entrance specific turbulence kinetic energy (ω), turbulence viscosity (v_t) and turbulence intensity (l_w) are derived from

$$\omega = \frac{\varepsilon}{C_\mu k} \tag{4.18}$$

$$v_t = \frac{k}{\omega} \tag{4.19}$$

$$l_w = \frac{\sqrt{\frac{2}{3}k}}{U(z)} \tag{4.20}$$

CFD models based on these equations, corroborated with instrumental data, show lower wind speeds at the toe of the frontal dune, accelerating up to the rim of the blowout. Once an erosional notch is formed, air flow is topographically steered into the throat of the blowout, often exceeding thresholds for sand deflation, generally modelled based on the Bagnold equation (or some modification thereof):

$$U_{*t} = A\sqrt{\left(\frac{\sigma - \rho}{\rho}\right) g\, d} \tag{4.21}$$

U_{*t} is the threshold shear velocity for dry sand, A is an empirical coefficient (=0.1), σ the density of the grain (typically 2650 km m^{-3} for quartz sand), and d the grain diameter (mm).

As sand is deflated from the center, the blowout deepens, creating a localized area of intense low pressure. This locally increases the pressure gradient, further accelerating the wind. Thus, the blowout initiation creates a positive feedback mechanism which tends to enlarge the feature until other factors become limiting.

"Greedy parabolic" dunes (crescent shaped dunes in vegetated areas with horns oriented upwind) direct wind within their deflation basins, governed by basin depth and width, in a manner phenomenologically similar to blowouts (Smyth et al., 2020). Deeper deflation basins and larger depth/width ratios are directly related to higher degrees of wind steering. Comparable positive feedback in process-form interactions also seems to occur in barchan dunes, though so far only CFD studies without field measurements have been conducted (Alvarez & Franklin, 2021).

4.1.4 Karst Breakthrough

The starting point for modeling evolving karst systems is a single isolated fracture. Constant hydraulic head drives calcite aggressive (i.e., capable of dissolving the surrounding rock) water from an input to an outlet. Nonlinear dissolution kinetics produce positive feedback whereby dissolutional enlargement at the outlet enhances flow. This further increases the widening, and so on until a threshold called a breakthrough event occurs. At this point flow increases dramatically, hydraulic head breaks down, and enlargement of the fracture is accelerated, but at a consistent rate along the entire length, under conditions of constant recharge. This is an example of gradient selection through the fracture, reinforced by the positive feedback of dissolution.

Dreybrodt and Gabrovsek (2003) showed that dissolution rates for limestone are given by

$$F_1 = k_1\left(1 - c / c_{eq}\right) \text{for}\, c < c_s \tag{4.22a}$$

$$F_n = k_1\left(1 - c / c_{eq}\right)^n \text{ for c} > \text{c}_\text{s} \tag{4.22b}$$

F *is* the detachment rate at mineral surface, and $k = \alpha\, c_{eq}$, where α is a chemical kinetic constant. Calcium concentration is c, with c_{eq} representing equilibrium concentration of Ca^{++} at saturation, and c_s is a reference concentration, $c_s < c_{eq}$, that varies from $0.7c_{eq}$ to $0.9c_{eq}$. The exponent n varies between 3 and 6.

In early stages flow in the aperture is laminar, and after short distances of flow from the entrance the solution comes close to equilibrium (i.e., $c/c_{eq} \rightarrow 1$). Close to equilibrium (c_s), calcite dissolution in carbonates is inhibited due to impurities in the limestone, and the dissolution rates fall by orders of magnitude to a nonlinear rate law.

As the aperture is widened flow becomes turbulent, and due to eddy mixing the concentration gradients are levelled out. The mixed bulk is separated from the rock surface by a diffusion boundary layer (DBL), the thickness of which depends on flow hydrodynamics:

$$\varepsilon = a / \frac{\left(\frac{f}{8}\right)(\mathrm{Re}-1000)Sc}{1+12.7\sqrt{\frac{f}{8}}\left(Sc^{\frac{2}{3}}-1\right)} = \varepsilon / Sh \tag{4.23}$$

Sh is the dimensionless Sherwood number $Sh = (kL)/D$, where k is a convective mass transfer coefficient, D a diffusion coefficient, and L a characteristic length. *Re* is the Reynolds number, a is aperture width, f is the friction factor, and *Sc* is the Schmidt number, which for water is treated as a constant, $Sc = 1000$. Resistance to mass transport in the DBL is also affected by conversion of CO_2. If the distance a CO_3^{2-} ion travels from the mineral surface before conversion to HCO_3^- (diffusion length) is small compared to DBL thickness ε, diffusion rate is limiting and the effective rate constant k falls with increasing ε. If the diffusion length is less than or equal to ε, CO_2 conversion becomes rate limiting, as it is chiefly influenced by the boundary layer.

Thus, the positive feedback that enlarges gradient-selected pathways in karst is strongly influenced by both physical and chemical dynamics and associated with a critical threshold. More details are given in Dreybrodt and Gabrovsek (2003).

4.1.5 The Job of the River

Geoscientists often conceptualize fluvial systems as geomorphic machines that shape the landscape. Rivers are seen as engines of mass flux, as conveyor belts moving weathered material from source to sink, and as sculptors of landforms. In the classic text *Fluvial Processes in Geomorphology,* Leopold et al. (1964) put it thusly: "Rivers are the gutters, down which flow the ruins of continents." Salisbury (1919) in *Physiography* was even more eloquent: "Rivers constantly are shifting solid matter from land to sea. This is, indeed, their great work. Even the water, which falls upon the land, but does not flow directly to the sea, helps to make the rock decay, and so prepares it for removal by running water. It may therefore be said that *every drop of water which falls on the land has for its mission the getting of the land into the sea*" (Salisbury, 1919, p. 104, emphasis in original). Mackin (1948) was more straightforward, stating that stream profiles are "organized to transport sediment."

Given their unquestionably critical role in shaping landscapes and transporting mass, it is hardly surprising that rivers have been perceived as sediment flux machines. "Rivers are the primary agents of erosion, transportation, and deposition in most landscapes . . . Over much of the Earth fluvial processes determine the overall form of the landscape" (Summerfield, 1991, p. 207). Fluvial drainage basins are the fundamental units of geomorphology, while sedimentary basins are the fundamental units of geology, Leeder (1997) maintained, with continent-to-sea sediment fluxes as the link between them. Walling (2006) argued that "the transfer of sediment from the land to the ocean by rivers must be seen as a key pathway of material transfer on Earth." This view of rivers as sediment transport entities is fundamental to concepts of steady-state and grade in fluvial systems. Those equilibrium notions are based on an approximate long-term balance between sediment availability and transport capacity, which remain a staple assumption of river and landscape evolution models to this day.

As a reaction to the entrenched idea that steady-state equilibrium sediment flux is a natural or normative condition in fluvial systems, rather than a reference condition or simplifying assumption that may be rarely present, in Phillips (2010; from which arguments in this section are largely drawn) I argued for a view of rivers as hydrological rather than geomorphological entities: The *job* of the river is to move water. The geomorphic *work* of the river, considerable though it may be, is a byproduct of the job of the river. The idea of *hydraulic selection,* a form of gradient selection, is key to this argument.

When more precipitation falls than plants can use or the soil can absorb and hold, it must go somewhere; it must be redistributed. This redistribution is driven by gravity, so the excess moisture goes down the topographic/gravitational gradient. Hydraulic selection comes into play as the water is redistributed: more efficient pathways persist and prevail; less efficient options go unused or dry up (e.g., Jerin et al., 2023). As channels develop through seemingly random local concentrations of water and shear stress (relative to local resistance), selection favors the steeper, smoother, and larger channels. In addition to gradient selection, hydraulic selection as described in Phillips (2010) also involves the development of networks, which will be discussed in Chap. 6.

The steepest gradients of total head guide are selected, or the so-called "path of least resistance." Flow resistance equations are of the form

$$f \propto RSV^{-2} \tag{4.24}$$

and for the Darcy-Weisbach equation

$$f = 8g\,RSV^{-2} \tag{4.25}$$

where f is friction factor indicating flow resistance, R is hydraulic radius (equal to cross-sectional area divided by wetted perimeter), and V is mean velocity. Discharge ($Q = w\ d\ V$) is imposed and w, d are the width of flow and mean depth. R is very close to d in many cases, as long at $w \geq 6d$. Thus, if Q is externally determined, $f \propto S$: the path of least resistance is the steepest path, consistent with experience and intuition. Water selects the steepest gradient available to it.

Total sediment transport capacity at a given location along a river is a function of cross-sectional stream power

$$\Omega = \gamma\,Q\,S, \tag{4.26}$$

where γ is the specific weight of water and usually treated as a constant, Q is discharge (m^3 s^{-1}) and S is friction slope, often approximated as channel slope. Steady state requires that $\Omega \approx \Omega_{cr}$, the threshold power to transport the available sediment load. With discharge externally imposed, the steady-state condition becomes $S = S_{cr}$.

When $\gamma QS > \Omega_{cr}$, $S > S_{cr}$ and excess energy must be dissipated (see Nanson & Huang, 2008). Such dissipation may occur in the vertical dimension (e.g., via incision or formation of riffles or steps and pools), or laterally, via bank erosion and channel widening or increased sinuosity. Where sinuosity increases or riffles/steps and pools form there is a net reduction in slope so that $S \rightarrow S_{cr}$. In this way the dissipation of excess energy leads to an adjustment toward equilibrium—but as an emergent byproduct, not because of any law or principle that moves rivers toward grade or steady-state.

If $\gamma QS < \Omega_{cr}$, $S < S_{cr}$. The sediment load cannot be fully transported, so the channel aggrades. The aggradation can sometimes displace flows and increase the likelihood of levee breaches and crevasses, some of which may result in avulsions or meander cutoffs. Avulsions always exploit cross-valley slopes steeper than the original channel slope, and cutoffs shorten the channel, increasing slope. The aggradation thus often leads to an increase in slope such that $S \rightarrow S_{cr}$ and again, an apparent adjustment toward equilibrium.

In the case of $S > S_{cr}$, the slope decrease is a byproduct of energy dissipation. Even if $S \approx S_{cr}$, this does not indicate a stable, persistent state, in that if the flow gains access to a steeper path, it will inevitably select it. Increased slope when $S < S_{cr}$ is made more likely by responses to channel aggradation. However, the imbalances ($S < S_{cr}$; $\Omega < \Omega_{cr}$) can persist for long periods if the necessary setup conditions for cutoff and avulsions are not present. Keep in mind that this gradient selection is highly localized—flowing water can only "see" and respond to conditions in the immediate vicinity (more on local gradient selection below).

The adjustments described above are common. Sea-level rise, for instance, may either increase or decrease slopes of the lower river reaches, depending on the relative effects of higher base level at the river mouth vs. shortening of length by sea-level encroachment. Steeper slopes may produce $S > S_{cr}$ and excess stream power. Dissipation of this energy must be primarily in the lateral dimension as the river is near base level, resulting in increased sinuosity, lowering slope gradients. A rising sea level example is the Trinity River, Texas (Phillips & Slattery, 2008), while a falling base level case is found in the lower Jordan River, a tributary to the Dead Sea (Hassan & Klein, 2002). Increased slope due to tectonic tilting can also result in slope adjustments via higher sinuosity (Ouchi, 1985; Schumm et al., 2000; Phillips, 2012). The formation of steps and riffles has been shown to be an energy dissipation response to excess transport capacity, especially in channels where lateral channel adjustments are limited (e.g. Marston, 1982; Chin, 2002, 2003; Richardson & Carling, 2021).

The slope adjustments when $S > S_{cr}$ or $S < S_{cr}$ are far from inevitable, however, and are affected by externally driven changes in sediment supply, discharge, or slope (e.g., tectonics), and by the pre-existing state of the fluvial system as aggradational, degradational, or steady state. When excess power is present, energy dissipation does not always reduce slope—dissipation can be accomplished with bed forms, other roughness elements, or channel width–depth ratios. Where slope is less than the critical value, aggrading systems can adjust slope via cutoffs and avulsions only where certain conditions exist. Generally, these are that channel aggradation exceeds floodplain aggradation, a potentially steeper cross-valley route exists, and trigger events produce crevasses. Furthermore, not all crevasses result in channel change (Slingerland & Smith, 2004; Phillips, 2009). Some possible pathways are summarized in Fig. 4.8.

The key aspects of the arguments in this section are:

- Water flows follow the available path of least resistance relative to gravitational forces driving flow, which usually means the steepest slope gradient.
- Surface flow systems exist to redistribute excess water, and owing to gradient selection, this takes the form of channelized flows.

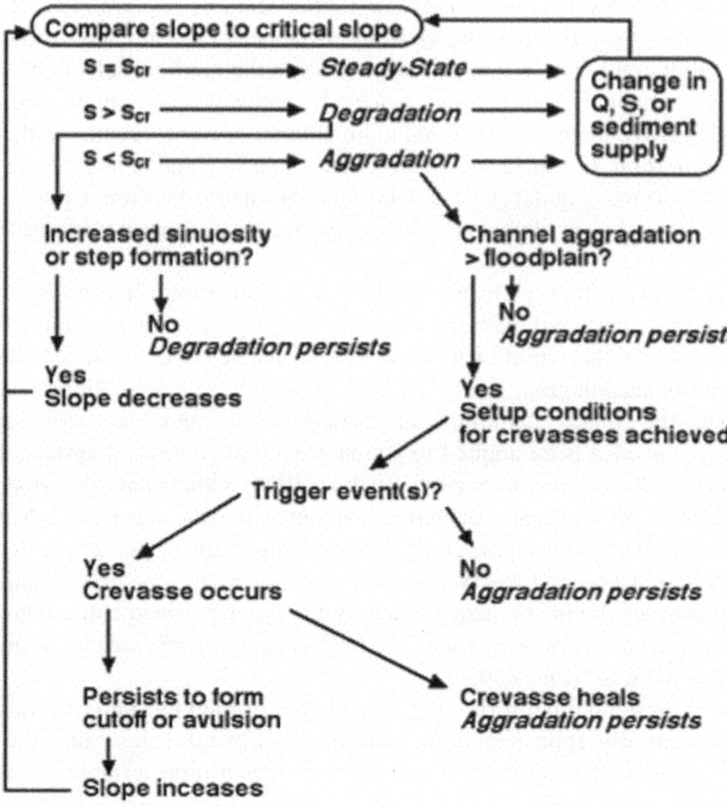

Fig. 4.8 Some possible fluvial system slope responses. (Reprinted with permission from Phillips (2009). © John Wiley & Sons, Ltd. All rights reserved)

- The geomorphic work of fluvial systems is a byproduct of the hydrological job of redistributing moisture, largely arising from dissipation of excess energy, or deposition due in inadequate energy to transport available sediment.
- Geomorphic adjustments of channels do not inevitably move the system closer to steady-state where sediment transport capacity matches sediment supply. However, sometimes this does occur as an emergent outcome, not as an equilibrium-seeking goal function.

4.2 Persistence and Recurrence of Preferential Pathways

Three different situations result in recurring or persisting preferential flow pathways. One is dictated by external factors, as discussed above in the context of wind. Some ocean currents and circulations channeled by sea floor bathymetry, or that are part of the general thermohaline circulation also fall into this category. A second is where flows encounter steeper gradients and are thereby directed or diverted along them. This is evident in soil and groundwater flows, where fractures, joints, root channels, soil cracks and other features capture flow. The third general situation occurs where flows along a steeper gradient carve a route such as a channel or conduit that persists and reinforces the gradient selection. In addition to fluvial channels, debris flows, avalanche pathways, and alpine glaciers are examples. Positive feedbacks that maintain, enlarge or reinforce the gradient flux paths are important. For instance, water flows may erode and enlarge a channel or maintain it by removing debris. The wetting of a surface facilitates flow, and the drowning of roughness elements reduces resistance and thereby increases flow velocity.

Once preferential flow paths are established canalization often occurs. This is when, once an evolutionary path is established, other previously possible pathways are eliminated, and the evolutionary trajectory is channeled, or canalized. The term was coined (by Waddington, 1942) and is most often used in evolutionary genetics (other than the literal definition of canal-building and river-straightening). Canalization has also been applied to geophysical and ecological systems (Caloi, 1964; Berlow, 1997; Levchenko, 1999; Stallins, 2006). Others have described canalized landform and landscape evolution without using the term (e.g., Naylor & Stephenson, 2010; Verleysdonk et al., 2011; Wainwright et al., 2011; Perron & Fagherrazi, 2012). Figure 4.4 is Waddington's metaphorical "epigenetic landscape," but well illustrates the phenomenon whereby the gravity-driven ball as shown has several possible pathways at the top of the landscape, but some are closed off as the ball moves downward (Fig. 4.9).

The location of channel formation may be determined by such tiny variations that they essentially appear random, as dynamical instabilities cause the minor change to grow. Once initiated, the new channel is a more efficient flow path, and those flows help maintain it and may enlarge it. In this example the channel location was not deterministically predictable, but once developed, other

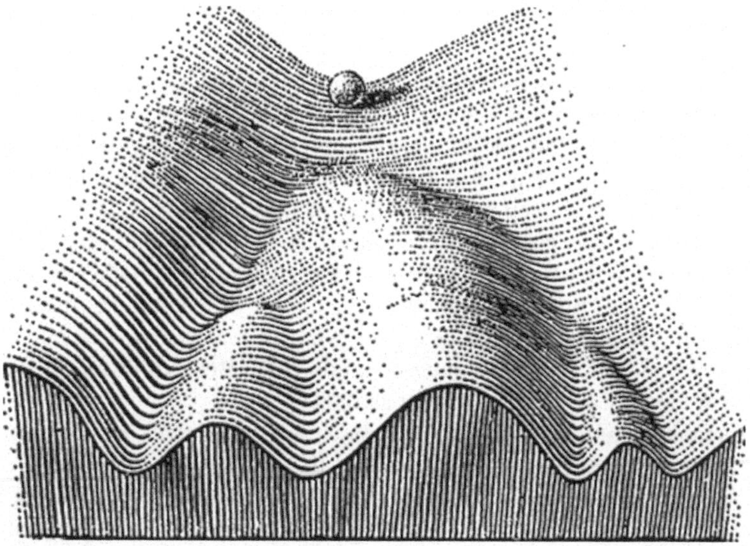

Fig. 4.9 Epigenetic landscape illustrating the idea of canalization. As the ball moves downhill, some initially possible pathways are excluded

previously possible channel locations are eliminated, and the developing channel strongly influences channel orientations downgradient. Dynamical instability and positive feedback often play a major role in canalization. Karst landform evolution is especially characterized by self-reinforcing positive feedbacks among dissolution, conduit and cavity enlargement, and hydrologic fluxes (Kaufmann, 2009).

Canalization often involves Irreversibility. Once a portion of a hillslope fails, for instance, the transported material cannot move back uphill. The mass movement and the changes in slope morphology caused by it affect future slope processes, making some future mass movements and other slope processes more, and others less, likely. Irreversibility is present in most weathering, erosion, and decomposition processes; canalization is therefore common.

4.3 Local Selection

Gradient selection is a prime example of the fact that selection is *local.* Though the aggregate effects of local selection throughout an ESS often enhance the efficiency of the system, that is not always so. Hillslope materials, ice, liquid water, and air have no awareness of what route is optimal for an entire hillslope or channel network—they can only "see" their immediate surroundings. A tumbling boulder has no way of knowing that a slight detour right or left will give them a quicker path downward.

4.3.1 Stream Capture

Jerin and Phillips (2017) demonstrated this in a fluviokarst system. A paleochannel was abandoned during downcutting of Shawnee Run, Kentucky, and the relocation was analyzed using a flow partitioning model. The paleochannel was a steeper, straighter, more efficient channel at the reach scale, but it was abandoned in favor of a longer, currently less steep and less efficient flow path. This occurred because Shawnee Run flow was captured by a subsurface karst flow path that was subsequently exhumed. When a swallet connected to a shaft opened in the stream bed the water could not "see" that it was relinquishing a more efficient route; it selected the local near-vertical gradient. Similar captures have been observed in other fluviokarst systems and are probably quite common (e.g., Becerril et al., 2010; Bocic et al., 2015; Phillips, 2017; Nogueira et al., 2023).

The general phenomenon of stream capture (also called river or stream piracy) is also an example of local gradient selection. In addition to subterranean capture, lateral planation or—most commonly—headward erosion of the capturing stream encroaches on the drainage basin of the captured stream. A steeper gradient is presented to the captured stream, whose flow is diverted to the channel of the pirate stream.

A recent example occurred in Yukon Territory, Canada, in 2016. Rapid climate-driven melting of the Kaskawulsh Glacier sent large quantities of meltwater into the Slims River, which formerly flowed into the Kluane River toward the Bering Sea. As overflow was diverted into the Kaskawulsh River, the locally steeper slope of the latter captured the Slims flow, diverting it toward the Pacific Ocean (Shugar et al., 2017). The Slims River quickly dried up as the regional drainage reorganization was driven by local gradient selection, independent (as is generally the case for piracy events) of broader-scale flow patterns.

4.3.2 Ice Pirates

An ice stream is a zone of faster-moving ice within an ice sheet or larger glacier, with velocities of up to 1000 m yr^{-1} or more. In Antarctica they are found mainly near the coastal margins and are critical features in land-ocean-atmosphere interactions. Glaciers move along gravitational gradients associated with the mass of ice overlying their base (i.e., ice thickness) and slope of the underlying topography. Their movement is also influenced by meltwater on, in, or beneath the ice. The latter is especially important due to its role in lubricating the ice/ground interface, reducing friction.

The Kamb Ice Stream (formerly known as Ice Stream C) in West Antarctica has effectively shut down for more than 150 years. The stagnation, atypical for other ice streams in the region, has been a subject of research since it was recognized in the late 1970s. Initial hypotheses focused on possible negative feedback mechanisms whereby rapid flow is counteracted by intervening episodes of stagnation or by a proposed feedback mechanism whereby ice flow speeds up until basal-water generation becomes sufficiently large that the water channelizes, lowers basal water pressures, and so slows or stops the ice (Rose, 1979; Retzlaff & Bentley, 1993).

Anandakrishnan and Alley (1997; see also Alley et al., 1994), however, argued that the slowing is restricted to the lower part of the glacier, caused mainly by a loss of meltwater lubrication of "sticky" spots on the ice stream bed. The resultant increased friction, they proposed, arises from local subglacial topography that directed water to a neighboring glacier. This apparent piracy of subglacial meltwater was due, in effect, to local gradient selection associated with what Anandakrishan and Alley called a "topographic accident," as small variations in bed topography can cause disproportionately large changes in ice velocity and water flow. Kim et al. (2016) studied a system of interconnected subglacial lakes beneath the Kamb ice stream, with results consistent with both the channelized basal water concept of Retzlaff and Bentley (1993) and water piracy. Kim et al. (2016) noted that both mechanisms may be operating. In 2021 a large meltwater channel comprising up to half the ice thickness was discovered near the grounding line of the ice stream, which was extending upstream and could be related to the proposed subglacial drainage reorganization (Whiteford et al., 2022). In another section of West Antarctica, Vaughn et al. (2008) found that the Carlson Inlet Glacier was not streaming, due to starvation of subglacial water due to piracy by the neighboring Rutford Ice Stream. They also found that the capture is sensitive to changes in ice thickness of <4%.

The dynamics of these and other ice streams, and subglacial hydrology more generally, are still under active investigation, but capture due to local gradient selection does occur.

Unlike surface water runoff, where gradients are determined by surface (land or ice) topography, basal meltwater hydrological gradients are strongly affected by thickness of the overlying ice. Assuming basal water pressure Φ is equal to ice overburden pressure (Patterson, 1994, chapter 6),

$$\phi = \rho_i g\left(z_s + 0.09 z_b\right) \tag{4.27}$$

A typical density of basal ice (ρ_i) is 917 kg m^{-3}, and z_s, z_b are the surface and bed elevations. Hydrological potential was calculated by Sharp et al. (1993) as

$$\phi = \rho_i g z_s + \left(\rho_w - \rho_i\right) g z_b \tag{4.28}$$

which is roughly equivalent to Eq. (4.27) at typical ice and water (ρ_w) densities.

In the modelling of subglacial hydrology by Drew and Tarasov (2023), the gradient of hydraulic potential is

$$\psi = \nabla\left(P_{water} + \rho_w g z_b\right) \tag{4.29}$$

where P_{water} is subglacial water pressure,

$$P_{water} = P_{ice}\left(\frac{h_{wb}}{h_c}\right)^{3.5} \tag{4.30}$$

P_{ice} is pressure due to the mass of overlying ice, h_{wb} is the basal water thickness, and h_c is thickness of saturated pore space.

These relationships highlight the sensitivity of subglacial flow to variations in ice thickness and subglacial topography, and to the thickness of the water layer vs. the thickness of the water accommodated within pore spaces. Drew and Tarasov (2023) confirmed the importance of subglacial hydrology in general and flow diversions in particular to glacial dynamics and mass balances, though they found that the details of process representation do not make much difference in predicting surge dynamics.

Paleo-ice sheet analysis also reveals evidence of capture-related reorganization of ice. On the Baffin Island shelf (Canada), Brouard and Lajeunesse (2019) examined a Pleistocene flow switch in an ice-stream due to glacial erosion of a marginal trough. They interpreted up-ice extension of ice streams through a marginal trough to have triggered capture of the neighboring ice catchment. This in turn provoked a flow switch and shutdown of an adjacent ice-stream. Their results suggest that competition for ice discharge between the ice streams (implying capture of ice catchments by marginal troughs) was the driver of ice flow switching. Ice flow switching and piracy has also been found for glaciers and ice sheets in Iceland (Benediktsson et al., 2022; Lally et al., 2024). Isolated topographic highs have the potential to locally slow or accelerate ice flow, as shown for the deglaciated landscape of the Cordilleran ice sheet in the northwestern U.S. (McKenzie et al., 2023).

The velocity of the surface of a glacier attributable to internal deformation is

$$u = k \sin^3 \alpha \, h^4 \tag{4.31}$$

where k is a constant, α the slope, and h the ice thickness. Since the underlying topography affects both slope and ice thickness, Eq. (4.31) indicates the large changes in velocity that could be associated with local variations in subglacial terrain (though McKenzie et al., 2023 found that effects of underlying topographic features are minimal below a size threshold).

The various forms of gradient selection discussed in this chapter are not independent of other types of selection. Gradient-driven fluxes encounter structures and substrates of varying degrees of strength and stability, preferentially eroding and removing the least resistant. This links gradient selection closely to resistance selection, which we turn to in the next chapter. The flow paths associated with gradient selection inevitably intersect to form networks, linking directly to principles of network selection (Chap. 6). Thermodynamic selection (Chap. 7) is largely based directly or indirectly on depletion of gradients as rapidly as possible.

References

Alley, R. B., Anandakrishnan, S., Bentley, C. R., & Lord, N. (1994). A water-piracy hypothesis for the stagnation of ice stream C, Antarctica. *Annals of Glaciology, 20*, 187–194.

Alvarez, C. A., & Franklin, E. M. (2021). Force distribution within a barchan dune. *Physics of Fluids, 33*, 013313. https://doi.org/10.1063/5.0033964

Anandakrishnan, S., & Alley, R. B. (1997). Stagnation of ice stream C, West Antarctica by water piracy. *Geophysical Research Letters, 24*, 265–268.

Becerril, J. A. O., Heydt, G. G., & Duran, J. J. (2010). The role of sculpted forms along endokarstic active conduits in the development of fluviokarstic canyons. The Rio Puron Cave conduit (Spain). In B. Andreo, F. Carrasco, J. J. Duran, & J. W. LaMoreaux (Eds.), *Advances in research in karst media* (pp. 387–392). Springer.

Bejan, A. (2007). Constructal theory of pattern formation. *Hydrology and Earth System Sciences, 11*, 753–768.

Benediktsson, I. Ö., Aradòttir, N., Ingólfsso, Ó., & Brynjólfsson, S. (2022). Cross-cutting palaeo-ice streams in NE-Iceland reveal Iceland ice sheet dynamics. *Geomorphology, 396*, 108009.

Berlow, E. L. (1997). From canalization to contingency: Historical effects in a successional rocky intertidal community. *Ecological Monographs, 67*, 435–460.

Bocic, N., Pahernik, M., & Mihevc, A. (2015). Geomorphological significance of the palaeodrainage on a karst plateau: The Una-Korana Plateau, Dinaric karst, Croatia. *Geomorphology, 247*, 55–65.

Brouard, E., & Lajeunesse, P. (2019). Ice-stream flow switching by up-ice propagation of instabilities along glacial marginal troughs. *The Cryosphere, 13*, 981–996.

Caloi, P. (1964). On the canalization of seismic energy. *Annals of Geophysics, 17*, 513–522.

Chin, A. (2002). The periodic nature of step-pool mountain streams. *American Journal of Science, 302*, 144–167.

Chin, A. (2003). The geomorphic significance of step-pools in mountain streams. *Geomorphology, 55*, 125–137.

Drew, M., & Tarasov, L. (2023). Surging of a Hudson Strait-scale ice stream: Subglacial hydrology matters but the process details mostly do not. *The Cryosphere, 17*, 5391–5415.

Dreybrodt, W., & Gabrovsek, F. (2003). Basic processes and mechanisms governing the evolution of karst. In F. Gabrovsek (Ed.), *Evolution of karst: From prekarst to cessation* (pp. 115–154). Zalozba ZRC.

Gares, P. A., & Nordstrom, K. F. (1995). A cyclic model of foredune blowout evolution for a leeward coast: Island Beach, New Jersey. *Annals of the Association of American Geographers, 85*, 1–20.

Hassan, M. A., & Klein, M. (2002). Fluvial adjustment of the lower Jordan River to a drop in the Dead Sea level. *Geomorphology, 45*, 21–33.

Hesp, P. A. (1996). Flow dynamics in a trough blowout. *Boundary Layer Meteorology, 77*, 305–330.

Hesp, P. A., & Hyde, R. (1996). Flow dynamics and geomorphology of a trough blowout. *Sedimentology, 43*, 505–525.

Hesp, P. A., & Pringle, A. (2001). Wind flow and topographic steering within a trough blowout. *Journal of Coastal Research Special Issue, 34*, 597–601.

Hugenholtz, C. H., & Wolfe, S. A. (2006). Morphodynamics and climate controls of two aeolian blowouts on the northern Great Plains, Canada. *Earth Surface Processes and Landforms, 31*, 1540–1557.

Hugenholtz, C. H., & Wolfe, S. A. (2009). Form–flow interactions of an aeolian saucer blowout. *Earth Surface Processes and Landforms, 34*, 919–928.

Jackson, D. W. T., Beyers, J. H. M., Lynch, K., et al. (2011). Investigation of three-dimensional wind flow behaviour over coastal dune morphology under offshore winds using computational fluid dynamics (CFD) and ultrasonic anemometry. *Earth Surface Processes and Landforms, 36*, 1113–1124. https://doi.org/10.1002/esp.2139

Jerin, T., & Phillips, J. D. (2017). Local efficiency in fluvial systems: Lessons from Icicle Bend. *Geomorphology, 282*, 119–130.

Jerin, T., Pavlowsky, R. T., Owen, M. R., et al. (2023). Anthropogenic channel network development in forested watersheds by recent stream channelization practices in Mark Twain National Forest, Missouri Ozarks. *Geomorphology, 426*, 108601.

Kaufmann, G. (2009). Modelling karst geomorphology on different time scales. *Geomorphology, 106*, 62–77.

Kim, B.-H., Lee, C.-K., Seo, K.-W., et al. (2016). Active subglacial lakes and channelized water flow beneath the Kamb Ice Stream. *The Cryosphere, 10*, 2971–2980.

Kleidon, A., Zehe, E., Ehret, U., & Scherer, U. (2013). Thermodynamics, maximum power, and the dynamics of preferential river flow structures at the continental scale. *Hydrology and Earth System Sciences, 17*, 225–221.

Lally, A., Ruffell, A., Newton, A. M. W., et al. (2024). Geomorphological signature of topographically controlled ice flow-switching at a glacier margin: Breiðamerkurjökull (Iceland) as a modern analogue for palaeo-ice sheets. *Geomorphology, 454*, 109184.

Leeder, M. R. (1997). Sedimentary basins: Tectonic recorders of sediment discharge from drainage catchments. *Earth Surface Processes and Landforms, 22*, 229–237.

Leopold, L. B., Wolman, M. G., & Miller, J. P. (1964). *Fluvial processes in geomorphology*. WH Freeman.

Levchenko, V. F. (1999). Evolution of life as improvement of management by energy flows. *International Journal of Computing Anticipatory Systems, 5*, 199–220.

Lin, H. (2010). Linking principles of soil formation and flow regimes. *Journal of Hydrology, 393*, 3–19.

Mackin, J. H. (1948). Concept of the graded river. *Geological Society of America Bulletin, 59*, 463–512.

Marston, R. A. (1982). The geomorphic significance of log steps in forest streams. *Annals of the Association of American Geographers, 72*, 99–108.

McKenzie, M. A., Miller, L. E., Slawson, J. S., et al. (2023). Differential impact of isolated topographic bumps on ice sheet flow and subglacial processes. *The Cryosphere, 17*, 2477–2486.

Nanson, G. C., & Huang, H. Q. (2008). Least action principle, equilibrium states, iterative adjustment and the stability of alluvial channels. *Earth Surface Processes and Landforms, 33*, 923–942.

Naylor, L. A., & Stephenson, W. J. (2010). On the role of discontinuities in mediating shore platform erosion. *Geomorphology, 114*, 89–100.

Nogueira, A. M., Caldeira, D., Uagoda, R., et al. (2023). Sedimentary records in dry karstic valleys: The case of Mambaí, Central Brazil. *Journal of South American Earth Sciences, 128*, 104338.

Ouchi, S. (1985). Response of alluvial rivers to slow active tectonic movement. *Geological Society of America Bulletin, 96*, 504–515.

Patterson, W. S. B. (1994). *The physics of glaciers*. Pergamon Press.

Perron, T., & Fagherrazi, S. (2012). The legacy of initial conditions in landscape evolution. *Earth Surface Processes and Landforms, 37*, 52–63.

Phillips, J. D. (2009). Avulsion regimes in Southeast Texas rivers. *Earth Surface Processes and Landforms, 34*, 75–87.

Phillips, J. D. (2010). The job of the river. *Earth Surface Processes and Landforms, 35*, 305–313.

Phillips, J. D. (2011). Emergence and pseudo-equilibrium in geomorphology. *Geomorphology, 132*, 319–326.

Phillips, J. D. (2012). Logjams and avulsions in the San Antonio River delta. *Earth Surface Processes and Landforms, 37*, 936–950.

Phillips, J. D. (2017). Landform transitions in a fluviokarst landscape. *Zeitchschrift für Geomorphologie, 61*(62), 109–122.

Phillips, J. D., & Slattery, M. C. (2008). Antecedent alluvial morphology and sea level controls on form-process transition zones in the lower Trinity River, Texas. *River Research and Applications, 24*, 293–309.

Retzlaff, R., & Bentley, C. R. (1993). Timing of stagnation of ice stream C, West Antarctica from short-pulse-radar studies of buried surface crevasses. *Journal of Glaciology, 39*, 553–561.

Richardson, K., & Carling, P. A. (2021). Morphology and origin of alluvial step-pools: A synthesis of experimental and field data from formative flows. *Earth-Science Reviews, 222*, 103823.

Rose, K. (1979). Characteristics of ice flow in Marie Byrd Land, Antarctica. *Journal of Glaciology, 24*, 63–74.

Salisbury, R. D. (1919). *Physiography* (3rd ed.). Henry Holt.

Samia, J., Temme, A., Bregt, A., et al. (2017). Do landslides follow landslides? Insights in path dependency from a multi-temporal landslide inventory. *Landslides, 14*, 547–558.

Schumm, S. A., Dumont, J. F., & Holbrook, J. M. (2000). *Active tectonics and Alluvial Rivers.* Cambridge University Press.

Sharp, M., Richards, K., Willis, I., et al. (1993). Geometry, bed topography and drainage system structure of the Haut Glacier d'Arolla, Switzerland. *Earth Surface Processes and Landforms, 18*, 557–571.

Shugar, D. H., Clague, J. J., Best, J. L., et al. (2017). River piracy and drainage basin reorganization led by climate-driven glacier retreat. *Nature Geoscience, 10*, 370–375.

Slingerland, R., & Smith, N. D. (2004). River avulsions and their deposits. *Annual Review of Earth and Planetary Sciences, 32*, 257–285.

Smith, A., Jackson, D. W. T., Cooper, J. A. G., et al. (2021). Whole-Island wind bifurcation and localized topographic steering: Impacts on aeolian dune dynamics. *Science of the Total Environment, 763*, 144444. https://doi.org/10.1016/j.scitotenv.2020.144444

Smyth, T. A. G., Jackson, D. W. T., & Cooper, J. A. G. (2013). Three-dimensional airflow patterns within a coastal trough-bowl blowout during fresh breeze to hurricane force winds. *Aeolian Research, 10*, 171. https://doi.org/10.1016/j.aeolia.2013.03.002

Smyth, T. A. G., Delgado-Fernandez, I., Jackson, D. W. T., et al. (2020). Greedy parabolics: Wind flow direction within the deflation basin of parabolic dunes is governed by deflation basin width and depth. *Progress in Physical Geography, 44*, 643–660. https://doi.org/10.1177/0309133319899306

Stallins, J. A. (2006). Geomorphology and ecology: Unifying themes for complex systems in biogeomorphology. *Geomorphology, 77*, 207–216.

Summerfield, M. A. (1991). *Global geomorphology*. Longman.

Twidale, C. R. (2004). River patterns and their meaning. *Earth-Science Reviews, 67*, 159–218.

Vaughn, D. G., Corr, H. F. J., Smith, A. M., et al. (2008). Flow-switching and water piracy between Rutford ice stream and Carlson Inlet, West Antarctica. *Journal of Glaciology, 54*, 41–48.

Verleysdonk, S., Krautblatter, M., & Dikau, R. (2011). Sensitivity and path dependence of mountain permafrost systems. *Geografiska Annaler A, 93*, 113–135.

Waddington, C. H. (1942). Canalization of development and the inheritance of acquired characteristics. *Nature, 150*, 563–565.

Wainwright, J., Turnbull, L., Ibrahim, T. G., et al. (2011). Linking environmental régimes, space and time: Interpretations of structural and functional connectivity. *Geomorphology, 126*, 387–404.

Walling, D. E. (2006). Human impact on land-ocean sediment transfer by the world's rivers. *Geomorphology, 79*, 192–216.

Whiteford, A., Horgan, H. J., Leong, W. J., & Forbes, M. (2022). Melting and refreezing in an ice shelf basal channel at the grounding line of the Kamb ice stream, West Antarctica. *Journal of Geophysical Research-Earth Surface, 127*, e2021JF006532.

Zhang, S., Qui, X., Fu, S., et al. (2022). Evolution of blowouts in artificial foredunes on Pingtan Island, China. *Earth Surface Processes and Landforms, 47*, 2597–2611. https://doi.org/10.1002/esp.5397

Resistance Selection

5

Abstract

Resistance selection is selection by persistence: forms, patterns, and interrelationships that last longer than their spatial or temporal neighbors are preferentially preserved. This can occur as the weakest and least resistant of existing materials are removed, or as positive selection for more stable and durable forms that emerge. Resistance selection is prominent in weathering, erosion and sediment transport by water and wind, and is also responsible for sorting and patterning in landscapes. Other examples discussed include soil and groundwater flow and evolution of coastal headlands.

5.1 Selection by Persistence

Forms, patterns, and interrelationships that last longer than their spatial or temporal neighbors are preferentially preserved. While this can occur due to sheer luck or happenstance, more often it happens because the preserved features are stronger, more durable, more stable, or in general more resistant or resilient. The general term for this selection-by-persistence is resistance selection. The most obvious examples are related to the chemical and physical resistance of rocks and geological materials to weathering and erosion. Weaker, more unstable, less resistant features are worn away more rapidly, leaving the resistance-selected, persistent features behind (Fig. 5.1).

A stability-based logical connection between the physical and biological worlds was proposed by Pascal and Pross (2016). Time and energy are the two facets of stability, they argued, and time, expressed as persistence, is the more general of the two. Their persistence principle "nature seeks persistent forms," is tautological but not wrong. In later work (Pross & Pascal, 2017) they developed the persistence principle in the context of thermodynamics of chemical kinetics. Dynamic kinetic

J. D. Phillips, *Abiotic Selection in Earth Surface Systems*, Geophysics and Environmental Physics, https://doi.org/10.1007/978-3-031-85862-8_5

Fig. 5.1 Resistant rock outcrops in Wadi al Hitan in the Egyptian Sahara

stability based on lifespan or persistence (rather than free energy), they propose, is the basis for understanding evolution. Two of the three universal concepts of selection proposed by Wong et al. (2023) are based on preferential preservation, and static and dynamic persistence (the third is novelty generation). The adaptive mechanisms identified by Egbert et al. (2023) in four non-biological chemical and physical systems are based on increasing the likelihood of their persistence.

In addition to selection by persistence, resistance selection in Earth surface systems can also be a form of gradient selection, whereby weathering and erosion proceed most rapidly along the least resistant avenues or within less resistant materials (Fig. 5.2). Thus, resistance selection can be seen as survival of the strongest or as removal of the weakest.

Resistance must be considered relative to the magnitude of the force or power involved, and to the process involved. Quartz sand grains, for instance, have a stable crystalline structure, low solubility, high strength, and are quite resistant to weathering. However, without some cementing or binding agent, sand has very low resistance to wind and water erosion. Carbonates (e.g. limestone) often have high strength and resistance to physical erosion, but high solubility and low resistance to dissolution. Depending on the setting, then, sand or limestone may be selected for preferential removal or preservation.

Resistance within the same material also changes, at various time scales. At geomorphological landscape evolution timescales, overall resistance tends to increase between clock-resetting-events (major events such as orogenies, volcanic eruptions, and major erosion or depositional events that "reset the clock" for landform development). This occurs due to the preferential removal of less-resistant materials and structures, increasing the proportion of more resistant ones. With respect to some

Fig. 5.2 The "sandstone city" at AdrŠpach-Teplice Rocks, Czech Republic. Water penetration and accelerated weathering and erosion along joints creates these spectacular passages. (Courtesy of Pavel Samonil. © 2018, Jonathan Phillips. All rights reserved)

processes, such as chemical weathering, resistance may change as weathering proceeds. In granite, for instance, low resistance biotites weather most rapidly, and resistant quartz most slowly, with feldspars intermediate. As weatherable minerals are depleted within a weathering profile, resistance thus increases. However, weathering also widens joints, fractures, and other partings, allowing more water access and decreasing resistance in that sense. These phenomena can lead to divergent weathering in early stages of rock breakdown (i.e., initial variations in resistance are magnified), and convergent weathering in later stages (Phillips, 2001; Fig. 5.3). Resistance can also change due to changes in vegetation cover or other biological agents, or due to armoring following selective removal.

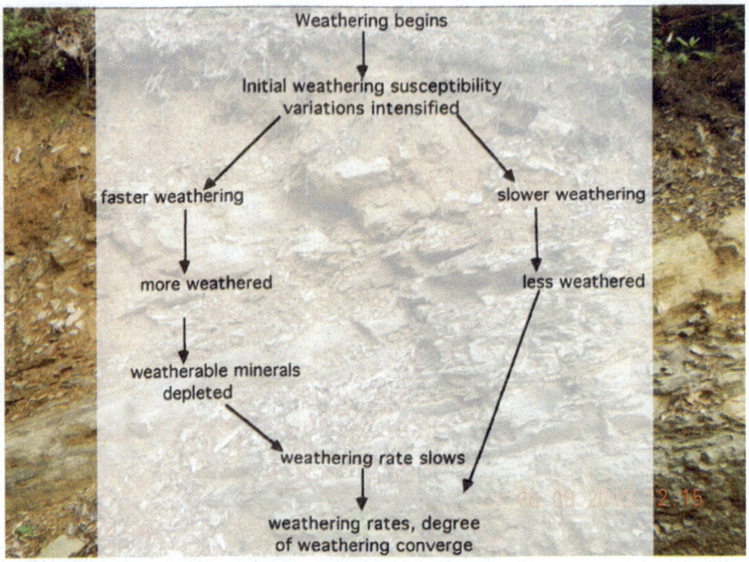

Fig. 5.3 Divergence of weathering rates and degree of weathering, followed by convergence (see Phillips, 2001). Background is weathering profile from Big Walker Mountain, Virginia

5.2 Force vs. Resistance in Erosion and Sediment Transport

Resistance selection in denudational processes is typically a threshold phenomenon that does not come into play unless the applied force exceeds the resistance. Weathering, erosion, and sediment transport are all controlled by force vs. resistance relationships. Resistance selection has long been a core idea in geomorphology.

For example, Sunamura (2018) developed a general equation for bedrock erosion by sediment-laden fluid flows of water and air:

$$\frac{d\Gamma}{dt} = C\frac{F_A}{F_R} - 1 \tag{5.1}$$

Γ is the amount of erosion, which in various cases might be measured in terms of eroded length, volume, or mass. F_A is the "assailing force" of sediment-charged water flows, waves, or wind, and F_R the resisting strength of the bedrock. The coefficient C has the same units as $d\Gamma/dt$. F_A is the product of fluid force and a dimensionless coefficient A, which reflects characteristic particle size. For instance, in the case of aeolian abrasion,

$$F_A = A(\textit{fluid force}) = A\,\rho_s\,u_s^2 \tag{5.2}$$

Where ρ_s is density of the entrained sediment (typically about 2.65 g cm^{-3} for quartz sand) and u_s is the velocity of sand particles impacting the rock surface. F_R

equals dimensionless coefficient B times an appropriate measure of resistance. In the aeolian example from Sunamura (2018),

$$F_R = B(resistance) = B\,S_c \tag{5.3}$$

with S_c taken as compressive strength of the bedrock. For this example, Γ is the depth of erosion relative to the original surface, z. The general Eq. (5.1) can be further modified to reflect specific circumstances. For example, as a depression excavated by aeolian abrasion deepens, the assailing force at the bottom decreases. Sunamura (2018) thus introduced an exponential decay function, producing

$$\frac{dz}{dt} = C\left[K\left(\rho_s u_s^2\right)/\,S_c\right]e^{-\beta z} - 1\} \tag{5.4}$$

β has dimensions of L^{-1} and represents the influence on depth of the assailing force. Equation (5.4) is a specific example of the general Eq. (5.1) using specific measures of erosion, assailing force, and resistance. Other examples are given by Sunamura (2018) for cases of abrasion by moving water in both fluvial and coastal settings.

5.2.1 Aeolian Sediment Transport

Much of the work on sediment transport by wind, to the present, represents applications, refinements, and extensions on the work of R.A. Bagnold in his 1941 book *The Physics of Blown Sand and Desert Dunes* (Bagnold 1941). In it, Bagnold recognized the importance of the threshold shear velocity required to entrain sand, u_{cr}. The critical shear velocity for a particular location is also an indication of the capacity of a surface to resist wind erosion:

$$u_{cr} = K_{w1}\sqrt{\frac{(\rho_s - \rho_a)}{\rho_a}\,g\,d_{50}} \tag{5.5}$$

K_{wl} is a constant (0.1 in Bagnold's experiments), and ρ_s and ρ_a are the densities of sand grains and air, with typical values of 2650 and 1.2 kg m^{-3}, respectively. The median grain size (diameter in mm), d_{50} is a commonly used representation of grain size.

Sediment transport rate as mass flow per unit width (Q, kg s^{-1} m^{-1}) is

$$Q = K_{w2}\sqrt{\frac{d_{50}}{d'_{50}}}\,\frac{\rho_a}{g}\,u^3 \tag{5.6}$$

where u is shear velocity, d'_{50} is a reference size of sand (taken as 0.25 mm), and the constant K_{w2} varies from 1.5 to 3.5 according to the grain size distribution. With minor differences, others followed the assumption of Bagnold regarding the shear velocity and threshold shear velocity, with all of them showing similar patterns when $u >> u_{cr}$ (Husemann et al., 2024). Many include an explicit representation of the force-resistance threshold, with no transport occurring unless $u - u_{cr} > 0$. For example, Kawamura's (1951) equation is

$$Q = K_{w3}\left(\frac{\rho_a}{g}\right)(u - u_{cr})(u + u_{cr})^2 \tag{5.7}$$

and Lettau and Lettau's (1978) is

$$Q = K_{w4}\sqrt{\frac{d_{50}}{d'_{50}}}\left(\frac{\rho_a}{g}\right)(u - u_{cr})u^2 \tag{5.8}$$

K_{w3} and K_{w4} were taken as 2.78 and 0.83. Kawamura's formulation incorporates a more detailed treatment of grain lifting and saltation, but like Bagnold's equation it simplifies the processes and does not fully represent the variability of wind conditions and surface properties. Lettau and Lettau extends Kawamura's work to apply to different environmental settings.

Resistance selection in aeolian processes explains several features of arid environments. Resistant bedrock residuals (yardangs) result from deflation of loose sediments and differential abrasion of rocks of varying resistance. Deflation of sand and silt often leaves a desert pavement surface of gravel atop cohesive soil (see Fig. 3.6).

5.2.2 Fluvial Sediment Transport

A standard approach to calculate initiation of motion of sediment in a fluid flow is the Shields (1936) parameter (also known as the Shields number or criterion):

$$\theta = \frac{\tau}{(\rho_s - \rho_w)g\,D_s} \tag{5.9}$$

Dimensional shear stress τ for fluvial processes is the product of specific gravity of water, depth, and energy grade slope ($\tau = \gamma\, d\, S$), ρ_s and ρ_w are the densities of sediment and water, and D_s is a characteristic particle diameter. θ is nondimensional and is proportional to the ratio of fluid force on the particle to the weight of the particle.

Bagnold also addressed fluvial sediment transport, and Bagnold (1980) produced a bedload sediment transport equation based on specific stream power (power per unit stream width ω, in W m^{-2}). Lammers and Bledsoe (2018) produced a refined version of Bagnold's equation:

$$q_b = \frac{a(\omega - \omega_{cr})^{1.5}}{D_s^{0.5}\, q^{0.5}} \tag{5.10}$$

Critical or threshold stream power necessary for entrainment ω_{cr} can be interpreted as an indication of bed material resistance. D_s is a representative grain size (in m, usually taken as D_{50}), q is unit water discharge (m^2 s^{-1}), and a is a constant (1.43×10^{-4}). The Lammers and Bledsoe equation is only one of many fluvial

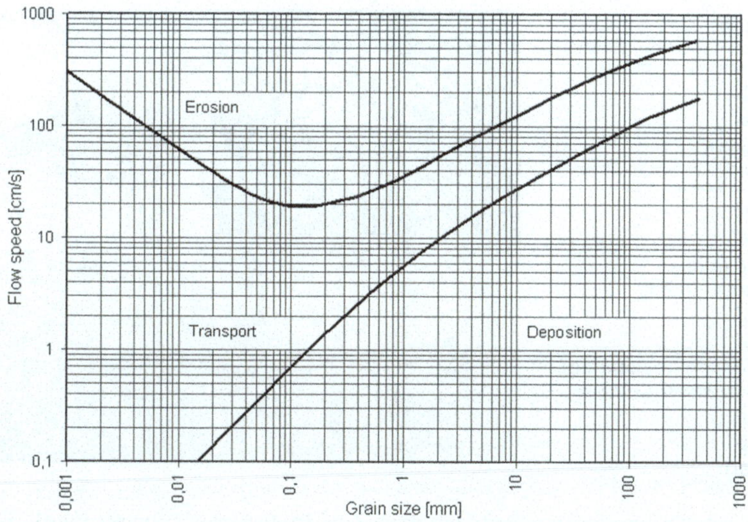

Fig. 5.4 The Hjulström curve. (Reprinted under CC-BY-SA-3.0 license from Wikimedia Commons. © 2009, The Author(s))

sediment transport relationships but is representative in the sense that it depends on exceedance of a power-resistance threshold for transport to occur.

Selective entrainment and transport are well illustrated by the Hjulström curve (Fig. 5.4), based on the experiments of Filip Hjulström (1935). The diagram shows the relationship between flow velocity and grain size. The upper erosion curve illustrates the higher resistance to entrainment by clay-size particles, where cohesion contributes to resistance. The threshold velocity decreases to a minimum in the silt size range, where cohesion is usually lacking, and increases again through sand and gravel particle sizes, where resistance is directly related to particle mass. The lower deposition line shows the threshold velocity to keep entrained particles in motion. Cohesion is not a factor in this case, and the smallest particles have settling velocities so low that any appreciable motion keeps them in suspension. Less force is required for transport than entrainment because grain interlocking and bed friction do not contribute to resistance for the former.

5.2.3 Sorting and Patterning

Force vs. resistance selection often produces sorting and patterning in landscapes associated with selective erosion, transport, and deposition, and resistance changes. Erosion is governed by the relationship between erosive forces (e.g., raindrop size, shear stress and stream power of flow, wind velocity) and resistance factors such as particle or aggregate size, cohesion, aggregate stability, shear strength, surface roughness, and vegetation cover. This results in selective, differential removal both within and between scales ranging from patches to drainage basins. Different

Fig. 5.5 Selective erosion, transport, and deposition. Photographic examples (clockwise from upper left) include sorting of riverbed sediments; erosion and deposition patterns in deserts; mosaics of water, tidal flat, marsh, and sandy or muddy sediments in coastal settings; and subaqueous bedforms

sediment sizes subjected to variations of velocities, discharges, slope gradients, etc. experience disparities in the frequency and distance of transport, as well as the mode (traction, saltation, or suspension). The same variations in force/resistance, along with buffers, barriers, and sedimentary accommodation spaces lead to uneven patterns of deposition. Deposited materials develop variable resistance based on time-in-place, vegetation establishment and other biological effects, and armoring or cementing. These selective processes always result in spatial variations, and sometimes create visibly evident patterning (Fig. 5.5).

Armoring occurs when smaller, lighter, less resistant material is removed, leaving behind more resistant sediment that shields less resistant material underneath. This is common in river bars and bedforms, and in some deserts, where the gravel lags form a surface called desert pavement. Vegetation establishment also increases resistance, as well as various forms of biofilms associated with bacteria, fungi, and algae.

5.3 Differential Weathering

Rock resistance to weathering can be measured in a variety of ways, including tensile, compressive, or shear strength, hardness, solubility, and chemical stability. Resistance is also influenced by rock partings—joints, fractures, bedding planes—which facilitate entry by water, roots, and other organisms, and affect the broad-scale strength and mechanical stability of rock masses and formations. Many formations not only have chemical and physical variations within a given lithology, but sedimentary rocks have multiple lithologies. The Ouachita Mountains (Arkansas and Oklahoma, USA), for example, include layers of weak, easily weatherable shales and siltstones, more resistant sandstones of variable composition and strength, and very strong cherts. These further occur with varying degrees of jointing and fracturing, in layers of various thickness, and which have been subjected to varying levels of tectonic compressive stresses (Fig. 5.6) (McFarland, 1998; Phillips & Marion, 2005; Von Voigtlander et al., 2018).

Exploitation of rock partings by water and biota, and more rapid weathering of less resistant materials results in exaggeration of initial variations in resistance and of local disturbances. This results in highly irregular weathering front geometry (Fig. 5.7) (Jones, 2013; Phillips et al., 2019; Šamonil et al., 2020).

Fig. 5.6 The author, along with Linda Martin and Alice Turkington, examining weathering features on a rock outcrop in the Ouachita Mountains, Arkansas. (Courtesy of L.L. Martin and A.V. Turkington. © 2005, Jonathan Phillips. All rights reserved)

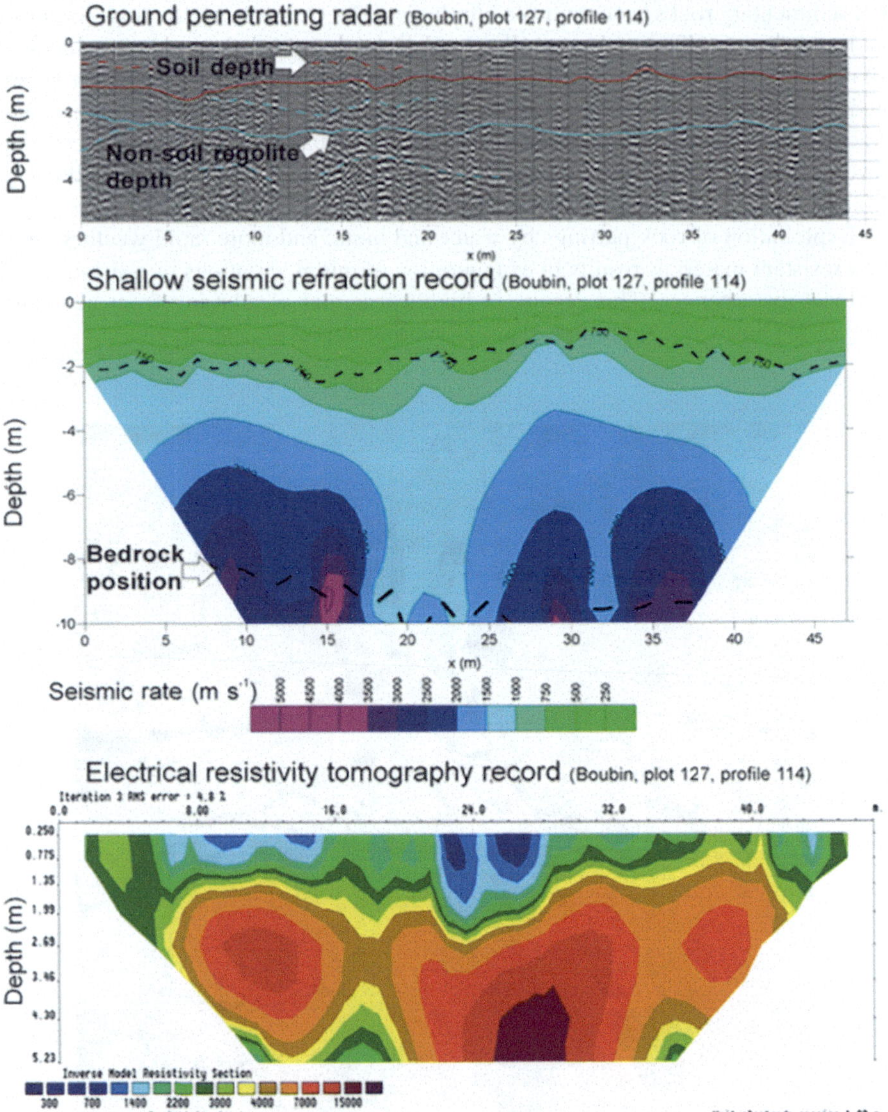

Fig. 5.7 Ground penetrating radar, seismic refraction, and electrical resistivity tomography representations of irregular weathering fronts from Boubin Reserve, Czech Republic. (Reprinted with permission from Šamonil et al. (2020).

5.4 Soil and Groundwater Flow

Resistance in ESS is also commonly encountered in the form of flow resistance. For flow through media or matrix material, resistance is the inverse of conductivity. Water flow through porous media such as soil or permeable rock is typically described by Darcy's Law,

$$V = K\left(h_2 - h_1\right)/d \tag{5.11}$$

V is velocity, *K* is hydraulic conductivity, and the latter term is essentially a slope gradient, reflecting the difference in hydraulic head *h* over distance *d*.

For a constant hydraulic head, *V* varies directly with *K*, which can differ even within apparently homogeneous materials such as sand, with variations in texture, composition, porosity, and antecedent water content. Faster movement along a wetting front into zones of higher conductivity (= lower resistance) can reinforce minor initial variations in *K*, leading to dynamical instability which can result in disproportionate exaggerations of even miniscule initial variations. This particularly comes into play when gravitational (as opposed to capillary) forces become dominant. Thus, the phenomena of unstable wetting fronts and fingered flow are an example of resistance selection (Fig. 5.8).

While wetting fronts are transitory features, dynamical instability and resistance selection of this type can result in persisting morphological changes (e.g., Ritsema et al., 1997; Montagne et al., 2013; Hermansen et al., 2019).

Fig. 5.8 Local variations in infiltration and wetting front movement in alluvial terrace soils in tropical Queensland, Australia

5.5 Coastal Headlands

Marine shorelines are often characterized by a range of headland-bay features with a promontory headland and associated embayment. These range from small, ephemeral beach cusps that may last less than one tidal cycle, to large, subcontinental scale capes. Headland-bay features are found on coasts in all types of materials, including sandy shores with minimal variations in resistance. On sandy shores the features are related to nearshore wave and circulation patterns, and to alongshore variations in sediment supply and removal. In some cases, these may reach a steady state form. For example, coastal engineers often use the parabolic bay shape equation of Hsu and Evans (1989; see also review by Hsu et al., 2010) to predict a stable equilibrium shape of bay beaches:

$$y = x\tan(\theta) + \frac{A}{\cos\theta} \tag{5.12}$$

where x,y respectively are the distance from the diffraction point along and perpendicular to the shoreline, θ is the characteristic or dominant angle of wave approach, and A is a constant that varies with beach geometry and wave conditions. The diffraction point is a promontory headland, structure (e.g., groin or breakwater), or similar feature.

On sandy beaches geological controls are often not evident, except where diffraction points are rocky features or where geologically controlled headlands are visible. However, the global review by Gallop et al. (2020) found that geologically controlled sandy beaches are the rule worldwide, not the exception (Fig. 5.9). Underlying geology influences erosional resistance, sediment supply and accommodation space; and rocky features can influence wave climates, wave energy dissipation, and coastal currents. They presented a conceptual model with a four-part typology of the degree of geological control, from slightly to deeply embayed. On such beaches resistance selection exists not only in the form of contrasting physical resistance of variable lithologies and shoreline materials, but also with respect to energy dissipation as waves approach shore, and sheltering from wave attack. Thus θ in Eq. (5.12), for example, is strongly influenced by nearshore bathymetry and may be closely related to basement geology.

Mean wave energy density per unit of horizontal area in J m^{-2} (per Airy linear wave theory) is

$$E = 0.5\rho g(H/2)^2 \tag{5.13}$$

where ρ is water density, g is the gravity constant, and H is wave height, equal to two times wave amplitude.

Wave energy at point i, at distance x_i from the shoreline is (Pruszak et al., 2008),

$$E_i = \frac{1}{8}\rho g\left(\frac{H_o}{h}\right)^2 A^2 x_i^{4/3} \tag{5.14}$$

Fig. 5.9 Sandtop Bay, Caldey Island, Wales, illustrating rock cliffs and headlands interspersed with beach embayments. (Photo by Ruth Sharville; reproduced from https://www.geograph.org.uk/photo/6864715 under CC BY-SA 2.0)

H_o is wave height entering the nearshore zone, h is water depth, and A is an empirical coefficient. Assuming energy entering the surf zone is E_o,

$$E_i / E_o = \left(\frac{H_o / h A x_i^{2/3}}{H_o} \right)^2 \tag{5.15}$$

From this it can be seen that for a given incoming wave height and energy, the E_i/E_o ratio decreases, and energy dissipation increases as the wave propagates into shallow water (lower h). Pruszak et al. (2008) applied the analysis to a multi-bar cross-shore profile of the Baltic Sea, showing the relationships among wave height, water depth, distance from the shoreline, and energy dissipation. The relative wave energy, as expected, decreases systematically with depth, which generally decreases with x_i, with superimposed variations from bars and other bathymetric irregularities. Using a more detailed representation of the effects of bottom friction and wave height variation, Chen et al. (2018) modelled the effects of complex bathymetry on wave energy distribution in Delaware Bay (USA).

Therefore, water depth and nearshore slope are an inverse form of resistance, selecting for specific shoreline features and configurations. On many headlands the associated bathymetry focuses wave energy on the headlands, with more dissipation and less energy in the embayments. The conventional wisdom was that this results in long term shoreline straightening due to preferential headland erosion and embayment accretion. However, the headlands often exist because of much greater physical resistance, which maintains the headland-bay form. Further, the promontories influence longshore sediment transport, affecting sediment supply up- and downdrift of the headland. For example, Short (2010) examined each of the 10,685 beaches of Australia to assess the degree of geological inheritance in the form of rocks, reefs, and headlands. Major effects included short mean beach lengths, reduction of breaker height in embayments, wave refraction, the presence of rip currents, and interruption or trapping of longshore sand transport.

Resistance and gradient selection are predominantly highly localized phenomena in ESS. We turn next to a generally broader scale, where selection favors flow and interaction networks in general, and specific network geometries and topologies in particular.

References

Bagnold, R. A. (1941). *The physics of blown sand and desert dunes*. Methuen.

Bagnold, R. A. (1980). An empirical correlation of bedload transport rates in flumes and natural rivers. *Proceedings of the Royal Society A, 372*, 453–473.

Chen, J.-L., Ralston, D. K., Geyer, W. R., et al. (2018). Wave generation, dissipation, and disequilibrium in an embayment with complex bathymetry. *Journal of Geophysical Research: Oceans, 123*, 7856–7876.

Egbert, M., Hanczyc, M. M., & Harvey, I. (2023). Behavior and the origin of organisms. *Origins of Life and Evolution of Biospheres, 53*, 87–112.

Gallop, S. L., Kennedy, D. M., Loureiro, C., et al. (2020). Geologically controlled sandy beaches: Their geomorphology, morphodynamics and classification. *Science of the Total Environment, 731*, 139123.

Hermansen, C., Moldrup, P., Müller, K., et al. (2019). Organic carbon content controls the severity of water repellency and the critical moisture level across New Zealand pasture soils. *Geoderma, 338*, 281–290.

Hjulström, F. (1935). Studies of the morphological activity of rivers as illustrated by the River Fyris. *Bulletin, Geological Institute Uppsala, 25*, 221–527.

Hsu, J. R. C., & Evans, C. (1989). Parabolic bay shapes and applications. *Proceedings of Institution of Civil Engineers, London, Part 2, 87*, 557–570.

Hsu, J. R. C., Yu, M. J., Lee, F. C., & Benedet, L. (2010). Static bay beach concept for scientists and engineers: A review. *Coastal Engineering, 57*, 76–91.

Husemann, P., Romão, F., Lima, M., et al. (2024). Review of the quantification of aeolian sediment transport in coastal areas. *Journal of Marine Science and Engineering, 12*, 755.

Jones, W. K. (2013). Physical structure of the epikarst. *Acta Carsologica, 42*(2–3), 311–314.

Kawamura, R. (1951). Study on sand movement by wind. *Report of the Institute of Science and Technology, University of Tokyo, 5*, 95–112.

Lammers, R. W., & Bledsoe, B. P. (2018). Parsimonious sediment transport equations based on Bagnold's stream power approach. *Earth Surface Processes and Landforms, 43*, 242–258.

Lettau, K., & Lettau, H. (1978). Experimental and micrometeorological field studies of dune migration. In H. H. Lettau & K. Lettau (Eds.), *Exploring the world's driest climate* (pp. 110–147). University Wisconsin.

McFarland, J. D. (1998). *Stratigraphic summary of Arkansas. Geological Commission Information Circular* (Vol. 36). Arkansas.

Montagne, S., Cousin, I., Josiére, O., & Cornu, S. (2013). Agricultural drainage-induced Albeluvisol evolution: A source of deterministic chaos. *Geoderma, 193*, 109–116.

Pascal, R., & Pross, A. (2016). The logic of life. *Origins of Life and Evolution of Biospheres, 46*, 507–513.

Phillips, J. D. (2001). Inherited versus acquired complexity in East Texas weathering profiles. *Geomorphology, 40*, 1–14.

Phillips, J. D., & Marion, D. A. (2005). Biomechanical effects, lithological variations, and local pedodiversity in some forest soils of Arkansas. *Geoderma, 124*, 73–89.

Phillips, J. D., Pawlik, L., & Samonil, P. (2019). Weathering fronts. *Earth-Science Reviews, 198*, 102295.

Pross, A., & Pascal, R. (2017). How and why kinetics, thermodynamics, and chemistry induce the logic of biological selection. *Bielstein Journal of Organic Chemistry, 13*, 665–674.

Pruszak, Z., Szmytkiewicz, P., Ostrowski, R., et al. (2008). Shallow-water wave energy dissipation in a multi-bar coastal zone. *Oceanologia, 50*, 43–58.

Ritsema, C. J., Dekker, L. W., van den Elsen, E. G. M., et al. (1997). Recurring fingered flow pathways in a water repellent sandy field soil. *Hydrology and Earth System Sciences, 1*, 777–786.

Šamonil, P., Phillips, J. D., Daněk, P., Beneš, V., & Pawlik, Ł. (2020). Soil, regolith, and weathered rock: Theoretical concepts and evolution in old-growth temperate forests, Central Europe. *Geoderma, 368*, 114261.

Shields, A. (1936). *Anwendung der Aehnlichkeitsmechanik und der Turbulenzforschung auf die Geschiebebewegung [Application of similarity mechanics and turbulence research on shear flow]. Mitteilungen der Preußischen Versuchsanstalt für Wasserbau (in German)* (Vol. 26). Preußische Versuchsanstalt für Wasserbau.

Short, A. D. (2010). Role of geological inheritance in Australian beach morphodynamics. *Coastal Engineering, 57*, 92–97.

Sunamura, T. (2018). A fundamental equation for describing the rate of bedrock erosion by sediment-laden flows in fluvial, coastal, and aeolian environments. *Earth Surface Processes and Landforms, 43*, 3022–3041.

Von Voigtlander, J., Clark, M. R., Zekkos, D., et al. (2018). Strong variation in weathering of layered rock maintains hillslope-scale strength under high precipitation. *Earth Surface Processes and Landforms, 43*, 1183–1194.

Wong, M. L., Cleland, C. E., Arend, D., Jr., et al. (2023). On the roles of function and selection in evolving systems. *Proceedings of the National Academy of Sciences (USA), 102*, e2320223120.

Network Selection

6

Abstract

Network selection in Earth surface systems (ESS) operates at two general levels. First, the emergence of interconnected networks of flows and interactions often supports the efficiency and survival of the systems. Second, in many cases specific network topologies are advantageous in terms of efficiency and are selected for. The first part of the chapter addresses the question of how strongly connected networks form in ESS, and their selective advantages are demonstrated using graph theory. An example of a highly connected fluvial-wetland complex and its ability to absorb impacts of extreme river floods and storm surges is presented. Percolation theory, derived from graph theory and statistical mechanics, is briefly summarized. The theory accurately predicts the emergence of optimal flux networks in soil and groundwater flow, the hydrological cycle, and other phenomena, illustrating the tendency for high-efficiency networks to be selected. Remarkable regularity in the branching dendritic topology of fluvial channel networks has long been noted and investigated. Channel networks differ in detail, but the statistical properties and the general visual impression are remarkably consistent across a range of spatial scales and environments. This is due to network selection linked to flux efficiency and energy dissipation. Network selection is also evident in rock fracture networks. Scale-free networks may also be subject to positive selection in ESS, but evidence is still limited.

6.1 Introduction

Some networks work better than others.

But what does "work better" mean? Speed or efficiency of movement or exchanges of information, mass, or energy? Stability? Redundancy? An effective tradeoff among two or more criteria?

J. D. Phillips, *Abiotic Selection in Earth Surface Systems*, Geophysics and Environmental Physics, https://doi.org/10.1007/978-3-031-85862-8_6

As you might expect, it depends on the network. But there are many cases in Earth surface systems (ESS) where networks work better than other means of accomplishing a specific task and where specific network characteristics, properties, or topologies are well suited to system persistence and efficient function. Such networks are preferentially selected.

As Paik and Kumar (2010) noted in their discussion of fluvial channel networks, "Mother Nature has left amazingly regular geomorphic patterns on the Earth's surface." That this occurs at a wide range of sizes and scales, in different climates, and in different geologic settings at least suggests a common underlying control, law, or principle. Paik and Kumar suggest that this is due to optimality according to the maximum entropy production principle, which will be covered in Chap. 7. Network selection provides a different take on this (and similar phenomena). Rather than a quest for optimality resulting in construction of particular types of networks, network selection is based on the idea that when patterns emerge (or are modified in ways that move an Earth surface system (ESS) closer to greater efficiency or stability), they are preserved and often reinforced. Patterns that do not increase stability and efficiency, or that work against it, are not preserved or reinforced, and may be destroyed or removed.

Like all selection, network selection is probabilistic. Further, while we assess network efficiency and stability at the scale of the entire network, the mechanisms by which networks are produced and modified (e.g., water flow, erosion, and deposition) are subject to gradient and resistance selection at a very local scale, as described in previous chapters.

We will start with a general discussion of why highly connected networks are typically advantageous to ESS, and therefore selected for.

6.2 Why Everything Is Connected to Everything Else

In a pedagogical or conceptual context in Earth and environmental sciences it is virtually axiomatic that everything is connected to everything else. That statement has been called the First Law—of ecology, of geography, and of environmental science. In ESS the first law (TFL) is generally true, although not all connections are direct, and some can be neglected in certain investigative context. In an article published in 2023, and on which this section is largely based, I addressed the question of *why* things become so densely interconnected. These inter-connections are readily treated as networks, so this question is fair game for the network selection chapter. Note that as compared to the more specific, concrete networks considered elsewhere in this chapter, based on networks of flow (e.g., hydrological networks) or physical patterns (e.g., rock fractures), here we consider a broad class of networks that includes flux networks and physical patterns, but also more general patterns of interaction among entities in environmental systems.

The *crowded landscape* concept recognizes that space and time are limited. Many different entities exist in ESS, some of which are constantly or intermittently in motion. With limited time and space for these to exist within, inevitably they

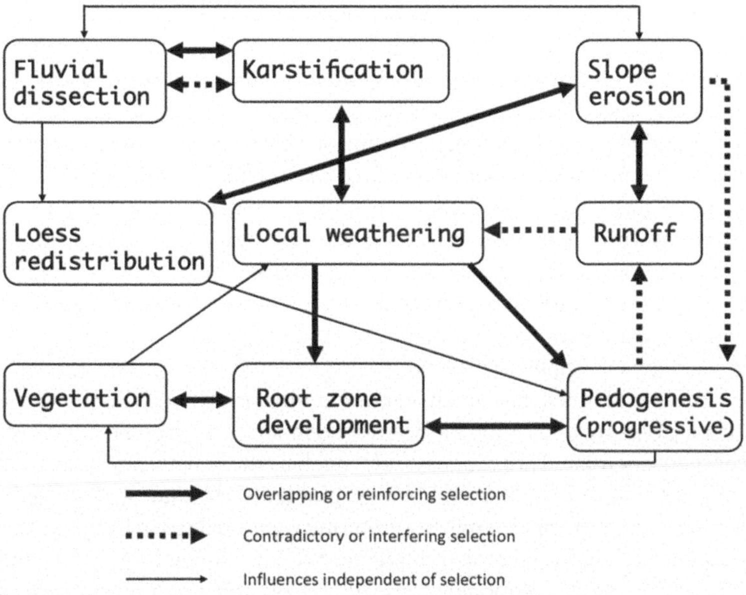

Fig. 6.1 Example of interconnections of various types among varied ESS components, based on studies of landscape evolution in fluviokarst areas of central Kentucky, U.S.A. (Reprinted with permission from Phillips (2023). © 2023 Elsevier B.V. All rights reserved)

encounter each other and often become connected in various ways. The *connection selection* concept is that there are advantages to connected ESS components that enhance their survival, persistence, recurrence and reproduction. Therefore, such connectedness is selected for. The crowded landscape and connection selection concepts are themselves strongly interrelated, as chance intersections in crowded landscapes are not likely to persist unless reinforced by positive feedback or selection.

In network terms, the elements or constituents of an ESS are referred to here as components, which correspond to nodes or vertices in graph theory. The connections between components are links, also called edges in graph theory. Components may be areas or locations, landscape features (e.g. landforms, soils, vegetation cover types, geological formations), system elements (e.g., source-transfer-sink zones, trophic levels, functional taxonomic groups), or process types or categories (e.g., erosion, deposition, evapotranspiration, infiltration) (Fig. 6.1).

6.2.1 Connectedness and Selection

6.2.1.1 Types of Connection in Earth Surface Systems

Connections commonly found in ESS and applicable to geophysical and other abiotic components or processes are summarized below.

Transfers and exchanges of energy, mass, and information are probably the most common type of connection. They can be bidirectional, as in reversing wind patterns or tidal fluxes, or one-way, as in gravity-driven flows. They can be reversible, as in downward or upward heat flux in soils as temperature gradients change, or irreversible, as in gravity-driven hillslope mass movements or runoff. Transfer links can involve transformations, as in state-changes of water, or not, as in the case of fluid or granular flows. Components connected by transfers can be sources, sinks, or pass-through, and these roles may be fixed or variable over time.

Competitive links are most evident in biology and ecology, but some abiotic processes are usefully seen and analyzed in terms of zero-sum competition, such as the allocation or budgeting of solar energy in climate, or of water in hydrological systems. *Spatial co-occurrence* connections are due to spatial adjacency of landscape elements, intersections of flux paths or structural features, or movement along the same pathways in opposite directions.

Links are also defined based on *mutual adjustments*, such as those between hydraulic variables in water flow, or between temperature, pressure, and volume in air masses. Closely related are *process links,* such as those between vegetation cover and erosion, or mass movements and slope gradients. *Mutual reinforcement* links between components are analogous to symbiotic relationships in biology; for example the relationships between moisture flux and water content in wetting from propagation. In some cases, components are linked by *self-limiting interactions*. For instance, floodplain or marsh surface elevations may be increased due to sedimentation during inundation, but as elevation increases inundation is limited. For a more extensive examination of the type of connections between components of ESS, see Phillips (2023).

6.2.1.2 Crowded Landscapes

Even a small patch of a terrestrial landscape is characterized by one or more landforms, soil or cover types, underlying geology, biological community, microclimate, hydrological status and function, etc. The patch will be the site of numerous movements into, out of, and through, which occur at different frequencies and intensities, and at, above, and below the ground surface, and in many directions. Water movements alone can include those driven by gravity, capillarity, root suction and other biological uptake, vapor transport, ice formation and melting, and state changes. In short, there is a lot going on, along many pathways (Fig. 6.2). Thus, innumerable prospects exist for entities and processes to encounter and to influence each other. Such encounters do not inevitably become persisting interrelationships, but the crowded landscape concept does provide necessary, though not sufficient, preconditions for high connectivity to emerge.

Additionally, there is a tendency for expansion. Hydrological flow networks often develop to become space-filling, expanding to the maximum density of channels that can be supported by runoff production. If progression of the weathering front exceeds erosional removal, soil and regolith profiles become deeper and thicker unless limited by the depth of landscape incision. This extends the more active zones of hydrological, biological, and chemical activity into the underlying rocks or sediments. Fluvial and dissolutional (e.g., karst) erosion generally involve progressive downcutting until limited by base levels. As is often assumed in

Fig. 6.2 Illustration based on the Sumava Mountains, Czech Republic, to show interactions among geology, soils, landforms, vegetation, and hydrology. Upward vertical fluxes in these landscapes include evaporation and transpiration, plant growth, soil mounding by plant growth, and tree uprooting. Downward vertical movements include infiltration and percolation, infilling of rock fractures and uprooting pits, leaf, litter, and trunk fall, and downward growth of roots. Lateral fluxes include surface runoff, erosion, and sediment transport; mass wasting; saturated throughflow and lateral groundwater flow, root growth along rock fractures, and aeolian transport of leaves and particulate organic matter. (Reprinted with permission.)

biogeographical and ecological models and theories, ecological systems tend to approach biological saturation, meaning that all available niches and habitats become occupied. All these expansionary trends create situations where different components and processes of ESS are increasingly likely to encroach or intersect. Regressive development and clock-resetting disturbances are sometimes present, but connectivity generally grows during expansionary development.

6.2.1.3 Connection Selection

This is the idea that some connections are advantageous, as they improve the odds of survival, reproduction, recurrence, and expansion of one or both connected components, or of an entire network or subnetwork. On average, connections that increase efficiency, stability, and resistance are preserved and sometimes enhanced. Equally important, happenstance connections that do not support stability, resistance, and efficiency are not preferentially selected, and are sometimes inhibited or eliminated.

For connections among or including biological entities this is well established. For individuals, connections that provide survival or reproductive advantages are more likely to be preserved than those that do not, via Darwinian natural selection. Selection also exists in the form of ecological filtering. Habitat qualities and resources can enable and encourage some and inhibit or discourage other organisms, and they can also promote or obstruct interrelationships. Connection selection is particularly important at the ecosystem level, as interrelationships such as those involved in energy and biogeochemical fluxes cycling are differentially selected for when they maximize efficiency and stability (e.g., Lotka, 1922; Lapenis, 2002; Verboom & Pate, 2013; Cong et al., 2017; Zhang et al., 2021).

Biogeochemical selection (see Sect. 3.1) occurs when more rapid (efficient) rates of energy and elemental cycling are preferentially formed and retained as ecosystems (and biospheres) evolve. Faster cycling provides advantages to individuals, but also to ecological systems as a whole—the more rapidly water nutrients and water are used and passed along, the more different entities can use them. Doolittle (2017), focusing on microbes, proposed that biogeochemical cycles are the unit of selection.

Abiotic examples of connection selection include hydraulic geometry, whereby stream channel width, depth, velocity, energy grade slope, and roughness (friction factor) all affect, and are affected by each other. This network of interactions is dynamically unstable over fluid dynamical time scales at a cross-section (Phillips, 1990, 1991), but this enables rapid, continuous adjustment to imposed flows and enhances stability at broader scales. A quick illustration is based on the Froude number, which distinguishes between subcritical and supercritical flow:

$$F_r = \frac{v}{\sqrt{gd}} \tag{6.1}$$

where v is flow velocity, g is the gravity constant, and d is water depth. Subcritical or tranquil flow ($F_r < 1$) is the norm in natural channels, with F_r often <0.5. In the subcritical regime the wave velocity exceeds the mean flow velocity and ripples can propagate upstream. $F_r > 1$ indicates supercritical flow. The large energy losses accompanying supercritical flow promote bank erosion and channel enlargement if shear stresses are sufficient. This in turn leads to a reduction in v and in F_r. Jia (1990) proposed a minimum Froude number hypothesis for alluvial channels, which is broadly consistent with the least action principle. Interactions between flow and channel morphology are also selected for stability, as extensively explored by

Huang and Nanson (2000, 2007, Nanson & Huang, 2008, 2017, 2018; see Sect. 1.3.2), among others.

Similar phenomena seem to occur with glaciers, via interactions among ice thickness, basal pressures and shear stresses, basal meltwater, and glacial movement. An increase in ice thickness increases the basal shear stress, with will accelerate the rate of ice flow and thereby locally reduce ice thickness. When the increased basal pressure causes melting, the meltwater lubricates the base, allowing for basal slip and faster ice movement. Shear stress at the base of the glacier is a function of ice thickness H and ground slope S:

$$\tau = \rho g H S \tag{6.2}$$

where ρ is ice density. Movement by plastic deformation is controlled by the strain rate e, as described by Glen's flow law as adapted by J.F. Nye for glaciers:

$$e = A\tau^n \tag{6.3}$$

A is a constant related to temperature and the exponent n is typically about 3, but ranges from 1.7 to 4.5. With $n = 3$, a doubling of effective shear stress results in an eightfold increase in strain rate. Occasional rapid movement responses—glacial surges—are generally attributed to basal slip resulting from the meltwater due to pressure melting in response to thicker ice. Frictional heating from the basal sliding provides a positive feedback to the basal pressure melting, further increasing movement. An alternative explanation for glacial surges is based on pore water pressure in subglacial sediments. When the latter are well-drained, pore water pressure is low, and deformation of the sediments (and associated ice movement) is nil. Increased basal shear stress, however, disrupts the drainage of the subglacial material, increasing pore water pressure and increasing deformation. Either way, the interconnections among ice thickness, shear stress, pressure melting, and glacial flow allow the ice to respond to changes in the ice mass. A clear summary of the basic principles is provided by Summerfield (1991: 264–269), and a more detailed and sophisticated approach to modeling glacier flow is given by Pelletier (2008: ch. 6).

Crowded landscapes ensure that connections are made, and connection selection that some are preserved and nurtured. But network, as opposed to connection selection, suggests some advantages to high connectivity for the system as whole.

6.2.2 Selective Advantages of Connections

More connectivity is clearly not always better—for instance, high connectivity can result in the rapid spread of misinformation, disease, pollutants, or computer viruses. In landscapes, however, connections are often beneficial, though Tiwari et al. (2024) show how increasing connectivity can either increase or decrease overall system efficiency in ecohydrological networks. Connections (see the typology in the next section) can be considered as exchanges of *information*. The state of, or changes in, one component can give indications of the other components connected to it. One definition of information is "what is conveyed by a particular arrangement or sequence of things" (*Oxford Dictionary of the English Language).* In the case of biological entities, information may be in the form of messages. Links among biotic components may allow a plant to sense an oncoming soil moisture deficit and communicate this information to other plants via mycorrhizal networks (e.g., Sheldrake, 2020). Several recent studies in hydrology have been based on transfers of information, in addition to fluxes of water and energy (e.g., Goodwell et al., 2018, 2020; Tennant et al., 2020; Li & Good, 2021).

In general, greater interconnectedness in networks has, in many situations, the following advantages:

- Faster, more efficient transfers and transformations of mass, energy, or information,
- More opportunities for partitioning of flow, and alternatives if a specific link is limited or unavailable.
- More options or degrees of freedom in adjusting or responding to disturbances or changes.
- Greater synchronization among system components.
- Reduced vulnerability. Strongly connected systems are more likely to survive and recover if a particular node is lost, and do not depend as heavily on critical nodes whose disruption or loss endangers the entire system.

The First Law (everything is connected to everything else) at broader, non-local scales is not always based on direct links, as many of the types of connections in ESS require spatial contiguity or proximity. Further, connectedness associated with, e.g., global biogeochemical cycles and ocean and atmosphere dynamics is often weak at broad scales. Sea surface temperatures or ice cover in one region of the ocean are connected to, e.g., precipitation or land temperatures far removed, but those connections are indirect and filtered. Thus, the arguments here apply at spatial and temporal scales at which the crowded landscape concept applies.

It is *not* the intent here to argue that ESS universally or even prevalently evolve toward fully-connected networks—after all, selection for tree-like and other network topologies is discussed later in the chapter. Rather the intent is to show that in a broad, general sense, more connectivity is often better and therefore selected for when the geophysical constraints allow it, which they sometimes do not. A soil pore at the surface of a soil profile, for instance, cannot become connected directly to one at the profile base without passing through other pores first.

6.2.3 Highly Connected Graphs

Consider an ESS as a graph with N components or nodes and m links or edges connecting them. Here we deal only with connected graphs, where every node is connected to at least one other and a path of some length exists between any two nodes. We consider only simple graphs, where any pair of components is either linked or not (1,0), notwithstanding the direction and number of the connections (for instance, land surface and troposphere conditions may be linked by insolation, albedo, precipitation, evapotranspiration, and other processes). Such a graph has a maximum of $N(N\text{-}1)/2$ links and a minimum $m = N—1$.

The graph's $N \times N$ adjacency matrix $\mathbf{A}$ has elements which are 1 if the row and column components are connected, and 0 otherwise, with zeros on the diagonal. Algebraic (or spectral) graph theory is based on analysis of $\mathbf{A}$, particularly its N eigenvalues λ_i, $\lambda_1 \geq \lambda_2 \geq \ldots \geq \lambda_N$.

The largest eigenvalue λ_1 is called the spectral radius, with a maximum value for a given N, m of

$$\lambda_{1,\max} = \left[2m(N-1)/N\right]^{0.5} \tag{6.4}$$

A fully connected graph—every node is directly linked to every other—has $\lambda_1 = N—1$. For a connected graph with the minimum m,

$$\lambda_{1,\min} = \left[(2N-2)/N\right]^{0.5} \tag{6.5}$$

The mean number of links per node for any graph is 2 m/N, so the upper bound can also be expressed as

$$\lambda_{1,\max} = \left[(mN-m)/N\right]^{0.5} \tag{6.6}$$

Spectral radius is commonly used as a measure of graph complexity. λ_1 increases with N, m, and the number of cycles within a graph, and is inversely related to the strength of interrelationships in networks of interacting time-varying systems, and thus related to coherence (Restrepo et al., 2006). Spectral radius is also closely related to various measures of graph entropy (Strydom et al., 2021; Geller et al., 2012; Mowshowitz & Dehmer, 2012).

Graph energy is the sum of the absolute values of the eigenvalues of $\mathbf{A}$:

$$E(g) = \Sigma|\lambda_i| \tag{6.7}$$

The concept and term arose in physical chemistry, where graph energy is used to estimate the total π-electron energy of molecules. Graph energy can be more generally interpreted as a representation of the total strength of feedbacks in the network, both positive and negative. In a simple undirected graph,

$$(2\mathrm{N}-2)^{0.5} < E(g) < (2\mathrm{N}-2). \tag{6.8}$$

The ratio of maximum to minimum possible graph energy is therefore $(2\,N—2)^{0.5}$. In a simple undirected graphs $E(g)$ is linearly related to λ_1:

$$E(g) = 2\lambda_1 \tag{6.9}$$

Algebraic connectivity α indicates synchronization of the network represented by the graph. It is measured by the largest non-zero eigenvalue (which is the second smallest λ) of the Laplacian of the graph adjacency matrix:

$$\alpha = \lambda(L)_{N-1} \tag{6.10}$$

Algebraic connectivity is constrained by

$$4/ND < \alpha < \kappa(G) \tag{6.11}$$

D is graph diameter (the minimum longest path between any two nodes) and κ(G) is vertex connectivity, the minimum number of nodes which could be removed to disconnect the graph. Maximum possible algebraic connectivity is associated with a fully connected graph, where $\alpha = N - 1$.

Algebraic connectivity also indicates the robustness of complex networks, because it measures the difficulty of separating the graph into different disconnected components; higher α values indicate more robust networks. Hembry et al. (2018) used algebraic connectivity as a proxy for propagation of ecological and evolutionary effects throughout a species assemblage. With higher α, effects such as convergence of traits or fluctuations in species abundance are more likely to affect multiple species within the network, due to interactions connecting different groups in the network. Analogous interpretations apply to geophysical networks.

Other ecological applications of algebraic connectivity are cited by Phillips (2023). Yazdani and Jeffrey (2012) and Phan et al. (2018, 2021) assessed robustness and redundancy of water distribution systems usings algebraic connectivity, which they found to be an effective measure of both, and higher α was considered unambiguously advantageous in that context. Algebraic connectivity has also been used to assess inferential synchronization of sedimentary and paleoenvironmental sequences and landform chronosequences (Phillips, 2013, 2016a, b).

The everything-is-connected-to-everything-else structure represented by a fully connected graph gives the highest possible spectral radius, graph energy, and algebraic connectivity for a given number of components. The fully connected configuration thus has the greatest graph complexity, the most total network feedback, and the highest synchronization, robustness, and connectivity. Compared to other system structures, a fully connected network can respond to change or disturbance with greater rapidity, synchronization, intensity, and with more degrees of freedom. Accordingly, higher connectivity is subject to positive selection to the extent that these represent advantages for network survival, growth, and perhaps propagation. Effects detrimental to network function or survival are also transferred faster and more extensively in more highly connected graphs, but even where this is the case it is often offset by more compensatory feedback.

Nonetheless, advantages exist for individual ESS components in sparser networks.

To receive or provide information, energy, or mass, or to facilitate quicker responses, it is advantageous for a component to be directly linked to other nodes. For a component to collect information (etc.) from multiple nodes a convergent radiation pattern is the most efficient. In a convergent radiation pattern a single key node is connected to every other node, which are connected only to the key node; Fig. 6.3). Maximum efficiency for transmitting or distributing to other components is a divergent radiation pattern. These structures are identical for undirected graphs, and for any *N*, $\lambda_1 = \sqrt{(N-1)}$.

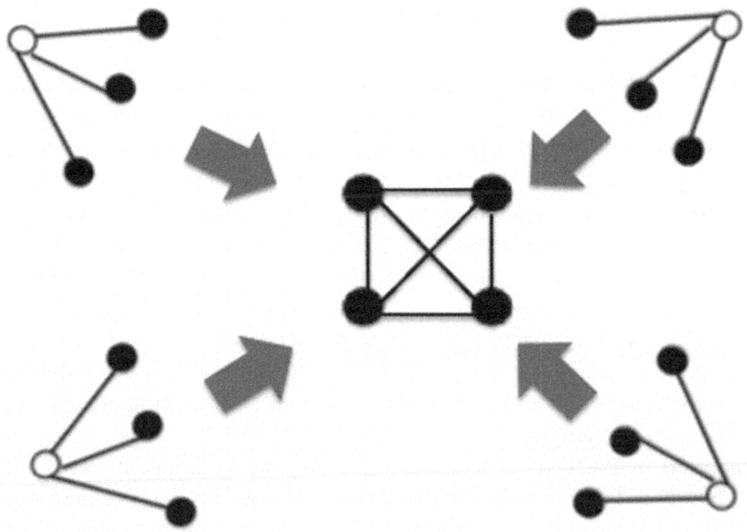

Fig. 6.3 Illustration of how four radiation graphs can be combined into a single fully connected graph. (Reprinted with permission from Phillips (2023).)

If a fully connected graph G is treated as a set of radiation subgraphs G_i (i = *1, 2,…, N*) we can apply a standard rule from graph theory,

$$\lambda_1(G) < \Sigma \lambda_1 G(i) \tag{6.12}$$

Where the subgraphs all have $N = 4$ nodes,

$$\Lambda = \Sigma \lambda_1 G(i) = 1.414N \tag{6.13}$$

If a fully connected ESS network is a collection of radiation subgraphs associated with each individual component,

$$\lambda_1 / \Lambda = (N-1)/1.414N \tag{6.14}$$

The fully connected structure as a collection of radiation subgraphs maximizes connectivity for both individual components and for the network but incorporates subnetworks for each component that minimize complexity (as indicated by λ_1) and also minimize complexity for the system relative to the summed spectral radii of the subgraphs.

As links m increase relative to N, synchronization also increases. Addition of a single link could leave graph diameter D unchanged but cannot reduce it. Further addition of links must at some point reduce D by reducing the maximum shortest path between any pair of nodes. This increases the lower bound of algebraic connectivity ($4/ND$; Eq. 6.11). Larger m must also increase the vertex connectivity, the upper limit for α.

6.2.4 Advantages of Connectivity: An Example

The channel-wetland complex of the lowermost Waccamaw River, South Carolina (Fig. 6.4), provides an example of the advantages of connectivity. In 4 years, the river experienced the three highest flows ever recorded (October 2015; October, 2016; and September 2018). The latter two events were associated with Hurricanes Matthew and Florence and included storm surges from the Winyah Bay estuary as well as river flooding from upstream. Notwithstanding those large, high-energy flows there were minimal changes in the hydrology, geomorphology, and ecology of the system—in sharp contrast to extensive impacts on humans and anthropic features. The ability of the system to absorb impacts is attributable to the connectivity among water conveyance and storage features, which allowed storage, delayed flow, and "spillway" conveyance of excess water, as well as activation, deactivation, and reversal of water exchanges.

As reported in Phillips (2023), hydrogeomorphic components were identified from fieldwork and aerial imagery (Table 6.1). Movements and exchanges of water among components were assessed using direct observations at varying river stages and field indicators such as sandy bedforms, flood wrack, and bent vegetation.

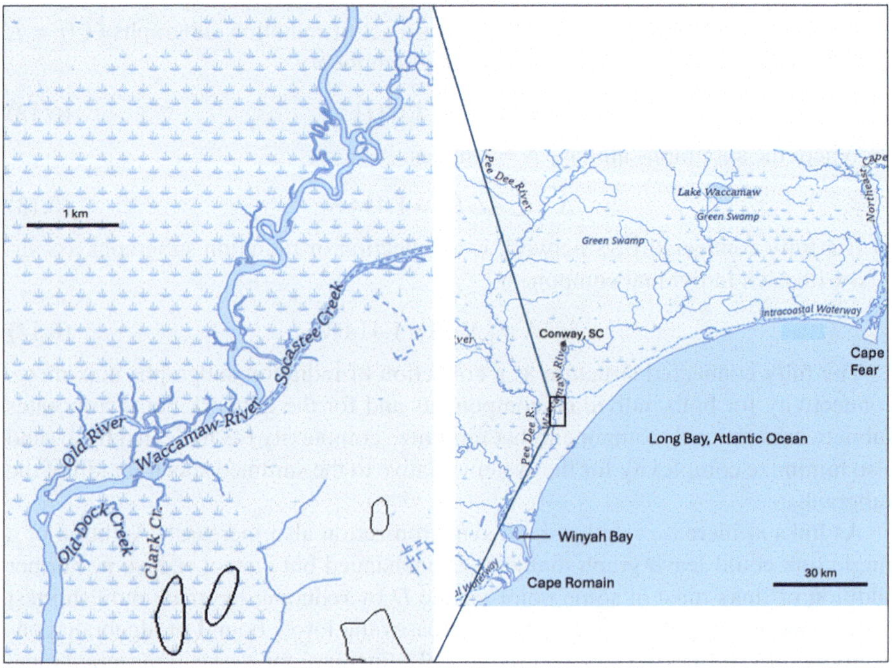

Fig. 6.4 Lower Waccamaw River, South Carolina. Inset shows example of channel/wetland complex. The paucity of contour lines and the swamp symbols indicate the low-relief nature of the system. Base map is from the U.S. Geological Survey National Hydrography dataset

Table 6.1 Hydrogeomorphic elements of the lower Waccamaw River. Descriptions refer to the features as they occur within the study area

Element	*Description*
Main channel	Dominant river channel.
Anabranch	Constantly flowing perennial channel roughly parallel to main channel and connected at both ends.
Tributary	Stream flowing into main channel or anabranch. In the study area all such confluences are within the valley bottom channel-floodplain complex.
Distributary	Streams and channels that normally direct flow from main channel or anabranches to floodplain basins or ACWBs (or further downstream, to estuaries). During falling limbs of floods, flow may temporarily reverse to return flow to channels. Distributaries are perennial; high-flow distributaries are considered flood channels.
Floodplain	Vegetated wetlands (mainly bottomland hardwood swamps) within the valley bottom that are seasonally to frequently flooded. These may convey flow in any direction during high stages, determined by river flows, astronomical tides, local runoff, storm surges, and wind.
Flood channels	Channels that are non-flowing or not inundated during normal or low flows that are activated to convey downstream or distributary flow at high stages.
Backwaters	Channels or ponded areas connected to main channel or anabranches that are normally backwater flooded, with ponded conditions or upstream flow.
ACWB (abandoned channel water bodies)	Floodplain lakes, oxbows, and sloughs hydraulically isolated from other channels during normal and low flows.

	MC	*Ana*	*Trib*	*Dist*	*FP*	*FC*	*BW*	*ACWB*
Main Channel		Flow split	Backwater	Flow split	Overbank flow	Flow split	Backwater	Overbank flow
Anabranch	Flow convergence		Backwater	Flow split	Overbank flow	Flow split	Backwater	Overbank flow
Tributary	Inflow	Inflow			Overbank flow	Inflow	Inflow	Overbank flow
Distributary	Return flow	Return flow			Inflow, overbank flow	Convergence		Inflow
Floodplain	Return flow	Return flow; cross flow	Return flow; local runoff	Cross flow; local runoff		Cross flow; local runoff	Return flow; local runoff; cross flow	Cross flow; local runoff
Flood channel	Convergence	Convergence	Backwater	Inflow	Inflow		Inflow	Inflow
Backwater	Inflow	Inflow	Backwater		Overbank flow	Backwater		
ACWB	Overflow	Overflow	Overflow	Overflow	Overflow	Overflow		

Fig. 6.5 Observed hydrologic connections and water exchanges among hydrogeomorphic elements of the lower Waccamaw River. Entries represent fluxes from the row to the column element. See text for explanation of terms

Connections among the hydrogeomorphic components are shown in Fig. 6.5. The connections are flows or exchanges of water and include *flow splits* (bifurcations of downstream channel flow) and *inflow* from external flows (tributaries) or exports from other components. *Convergence* indicates anabranches or distributaries joining or rejoining other channels. *Backwater* effects impede downstream flow or force water upstream. *Cross flow* occurs across floodplains; *return flow* returns water from floodplains and distributaries to channels. *Local runoff* includes water table rise within the floodplain, groundwater flow, and surface runoff. *Overbank flow* spills laterally out of channels, and *overflow* refers to drainage from filled ACWBs to other elements. Figure 6.5 applies to general components of the system, though specific features in lower Waccamaw are connected as shown. For instance, all flood channels are connected to the main channel, at least one anabranch and tributary, a distributary, floodplains, backwater channels, and ACWBs, as shown. But *all* flood channels are not connected to *all* the other individual elements.

Note that when two components are linked (and most are), in all instances connections are two-way, the net direction of flux varying with river stages and whether stages are rising or falling, and are also influenced by storm surges, tides, and wind. Figure 6.6 shows the connections as a simple, undirected graph with a spectral radius of 6.335, graph energy of 12.670, and algebraic connectivity of 6.000. These have maximum possible values of $\lambda_1 = 7$, $E(g) = 14$, and $\alpha = 7$ (for N = 8). Actual values are about 90% of the maxima, indicating high complexity and connectivity, strong feedbacks, robustness, and synchronization.

The high connectivity is very beneficial for handling high water impacts in a humid, low-lying system subject to tropical and extratropical cyclones, and other river/estuary systems with intact channel/wetland complexes are similarly advantaged (Phillips, 2022a, 2022b).

A counterexample in the same region shows how limited connectivity makes a wetland system vulnerable to change. Along the valley sides of the Neuse River estuary, North Carolina, about 240 km northeast of the lower Waccamaw, fluvially dissected valley sides occur. Incised during lower sea-levels in the Pleistocene, these short freshwater tributaries terminate at the estuary shoreline as ravine swamps. They are mainly fed by freshwater runoff, with occasional inputs from estuarine storm surges.

Hydrogeomorphic components are the incoming channel, a ponded swamp, one or more perennial outlet channels, a drip line at the downstream edge of the swamp

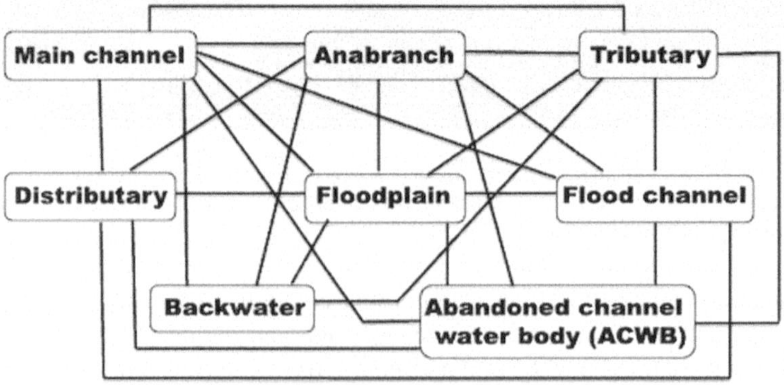

Fig. 6.6 Connectivity graph for hydrogeomorphic elements and water exchanges in Fig. 6.5

where diffuse discharge occurs during high water periods, a sandy beach (usually <10 m wide), and the estuary. Under normal, non-storm conditions there is net downstream flow from the incoming channel to the swamp to the out-flowing channel and across the beach to the estuary. Flow across the beach may be a single channel, multiple channels, fan-like diffuse flow, or a combination, depending on water level and wave conditions in the estuary. A graph reflecting these connections is below.

Graph metrics for the ravine swamp network are $\lambda_1 = 2.236$, $E(g) = 4.472$, and $\alpha = 0.586$ when the dashed link of Fig. 6.7 associated with storm surge is excluded, and 2.558, 5.116, and 0.914, respectively, with the storm surge link included. These values are considerably lower than the maxima for $N = 6$ systems—about 45 to 51% for spectral radius and graph energy, and about 12 to 18% for algebraic connectivity. This contrasts with the Waccamaw, where values are about 90% of the maxima.

The Waccamaw (and the analogous fluvial-estuarine transition zone of the Neuse River upstream of the ravine swamps) absorbed Hurricane Florence's impacts with minimal change. The ravine swamps, by contrast, were overwhelmed. The ponded swamps were inundated by estuarine storm surge for several days, which overwhelmed the spillway and secondary storage capacities. Portions of the ravine swamps were transformed from wetlands with perennial standing water and muck or mucky clay substrate to rarely flooded sandy lowlands (Phillips, 2022a). A channel as primary outlet with driplines activated during high flows was converted to the reverse at one field site, with driplines as the normal outflow and channels flowing

Fig. 6.7 Ravine swamp, Neuse River estuary. Photo on left shows outlet channel and beach; the swamp is the wooded area in the background. Diagram on right shows major components and their connections. The dashed line shows a connection that exists only during storm conditions when the swamp, its normal outlets, and the beach are inundated by storm surge. (Reprinted with permission from Phillips (2023).)

only during wet periods. At another site <500 m away the reverse transition occurred as a dripline outfall was transformed to a channel outlet.

The ravine swamp case shows how a low-connectivity system is less resistant and resilient to change, and thus not subject to positive selection.

6.2.5 Advantages of Connectivity: Summary

Selection seems to favor, on balance, highly connected configurations—a form of network selection.

In ESS connectivity emerges and increases because of the crowded landscape and connection selection principles. Processes operate continuously, transporting and transforming mass, energy, and information. Their operations inevitably encroach on each other, and transport pathways intersect, creating connections among the components involved. Those connections are often preserved and reinforced by positive feedbacks, mutually beneficial effects on interacting components, the irreversible nature of some processes and pathways, and efficiency advantages of using already-established pathways.

Connectivity between ESS components frequently produces net benefits to the components. Higher connectivity is associated with more intense and rapid energy and material cycling, more degrees of freedom for adapting to change, and greater synchronization of evolution and responses, thereby bestowing benefits to the network/system. Though higher connectedness can sometimes be deleterious, disadvantages are often offset by the benefits and the simultaneous achievement of maximum network efficiency for individual components and reduced cumulative complexity for the whole system.

With respect to hydrological and geomorphological phenomena, preservation and enhancement of advantageous connections is often achieved by processes associated with gradient, resistance, and efficiency selection. Steeper flux gradients, landscape elements with the higher resistance to weathering and erosion, and configurations with the greatest efficiency for work (according to the least action principle) preferentially occur, survive, recur, and grow (Hunt, 1998, 2017; Huang & Nanson, 2000; Phillips, 2010, 2011; Smith, 2010; Nanson & Huang, 2017, 2018). For ecosystems and intertwined biotic and abiotic phenomena, the advantages of strong connectedness are outlined by, e.g., Smith (1986), Lapenis (2002), Lekevicius (2002, 2006), van Nuland et al. (2016) and Ware et al. (2019).

6.3 Percolation Theory

Percolation theory is closely related to network analysis. It has been widely used in statistical mechanics, but as the name implies it is applicable to hydrological and geophysical problems. In simple terms, percolation theory describes the behavior of networks when nodes or links are added, reflecting a geometric type of phase transition. If you pour a liquid on porous material, can it find a path from hole to hole and flow to the bottom? Mathematically this is represented as a network of sites or vertices with bonds or edges. These can be open, allowing liquid in, with probability p. Or they can be no-flow (closed), with probability 1-p. It would be possible to have a series of open sites adjacent to each other along the flow gradient. But percolation theory is based on random open/closed patterns. For a given network (think of a 2-D or 3-D lattice structure), for a given p, what is the probability of an open path existing from one end to the other? Alternatively, what is the critical probability p_c at which there must exist an open path (recognizing that in some cases a path may exist when $p < p_c$). In this context you can think of 2-D as a cellular grid, with the cells (sites) open or not, and whether flow could (say) enter a cell on the left side and eventually exit on the right. 3-D is readily conceptualized as a rectangular column of cells, with flow moving from top to bottom.

Berkowitz and Balberg (1993: 775), in their overview of percolation theory applications in groundwater hydrology, state that "the principal advantage of percolation theory is that it provides universal laws which determine the geometrical and physical properties of a system. These laws are independent of the local geometry or configuration of the system."

Percolation theory assumes that the medium is infinite, with statistical properties that do not vary with location, and the occupation probabilities are random and independent. These assumptions are obviously not true for soils or geological media, though at the pore scale the extension of results to infinity is not unreasonable from an analytical perspective (Berkowitz & Ewing, 1998). Layering and horizonation are the most obvious, but not the only, sources of non-stationarity. The extent to which percolation theory is successful in describing or predicting real-world phenomena does not necessarily have any bearing on the relevance of the assumptions with respect the nature of the media. Rather, it indicates whether optimal pathways and networks tend to be selected in flux processes.

For groundwater flow, for example, water flow volume Q is related to the number of open sites N, with the number at the threshold (p_c) indicated by N_c.

$$Q \propto (N - N_c)^{\kappa} \tag{6.15}$$

The exponent κ can be theoretically derived or empirically estimated and is one example of numerous power laws in percolation theory where the exponents are constant and universal for either 2- or 3-D patterns. For instance, P is the probability that a given site is occupied and belongs to a finite cluster,

$$P \propto (p - p_c)^{\beta} \tag{6.16}$$

For a 2-D lattice, p_c is approximately 0.593.

Because of this, the fractal dimension of the percolation backbones (see below) is known for various network types.

A group of connected sites is called a cluster, and at larger p there is an increased probability of finding larger clusters. The cluster associated with the end-to-end path of open sites is known as the percolation cluster and is associated with the probability p_c. The set of system elements through which fluid can flow is called the "backbone" of the system. The path described by the backbone is fractal, with a dimension $D_b = 1.87$ for flow through a 3-D field. Though this value is not universal, Hunt et al. (2024) characterized it as quasi-universal, with the 1.87 value applicable for drying or saturated conditions.

Percolation theory has been successfully used to model and predict unsaturated hydraulic conductivity (permeability) in dual-porosity agricultural soils (Ghanbarian, 2021), permeability of rock fracture networks (Jafari & Babadagli, 2013; see rock fracture section below), plant root growth and the topology of root networks (Hunt, 2016), river basin and channel network morphology (Hunt et al., 2023), and solute transport (Hunt & Ghanbarian, 2016; Hunt, 2016). Allen G. Hunt and colleagues have also applied percolation theory to explore linkages among processes such as soil and groundwater flow, solute transport, soil formation, ecosystem productivity, evapotranspiration, and hydrological system organization (Hunt, 2016, 2024; Egli et al., 2018; Prabhakaran et al., 2021a; 2021b).

Two recurring elements in this work are: (1) the ability to predict hydraulic conductivity K for porous media based on the assumption that flow is optimized to minimize overall flow resistance (in essence the inverse of K); and (2) fundamental scaling relationships.

Overall K in a two-dimensional system follows a universal power law relationship:

$$K \propto \left(\frac{N}{N_c} \right)^{1.3} \tag{6.17}$$

and for 3-D

$$K \propto \left(\frac{N}{N_c} \right)^{2.0} \tag{6.18}$$

Transport time t through the system is proportional to a power of system size x,

$$t = t_o \left(\frac{x}{x_o} \right)^{D_b} \tag{6.19}$$

D_b is the fractal dimension of the percolation backbone. The characteristic length and time scales (x_o, t_o) are taken as pore size and 1 sec, respectively in Hunt (2016). The key scaling relationships are in the general form of Eq. (6.19); e.g., used to define optimal paths through soil:

$$x = x_o \left(\frac{t}{t_o} \right)^{1/D_o} \tag{6.20}$$

where D_o is the tortuosity exponent for an optimal path (1.21 for 2-D; 1.43 for 3-D).

This work is based on a tendency for water to follow paths of least resistance—sometimes phrased in terms of water "seeking" such paths, with the scare quotes sometimes used in the relevant papers (e.g., Hunt et al., 2023: 12). In terms of its physical manifestations, this is consistent with the idea that the path of least resistance is that most likely to be utilized, and that such optimal paths are subject to positive selection. The paths defined by percolation theory are those that minimize flow resistance and maximize conductivity (and are thus consistent with constructal theory). Comparison of the percolation theory predictions with large global datasets reveals a high degree of agreement (e.g., Hunt, 2016, 2017, 2024; Hunt et al., 2021, 2023).

This confirms the notion of network selection with respect to water flows, and to related phenomena such as solute transport, root growth, weathering, soil formation, and drainage basin evolution.

6.4 Dendritic Flow Networks

Remarkable regularity in the topology of fluvial channel networks has long been noted and investigated. Channel networks differ in detail, but the statistical properties and the general visual impression are remarkably consistent across a range of spatial scales and environments (Fig. 6.8).

Some models of channel network and drainage basin evolution rely on least work principles, whereby the network evolves to minimize the energy utilized to perform work or some related quantity or maximize thermodynamic entropy. These ideas have roots that go back to the 1960s and early 1970s. Leopold and Langbein (1962) argued that river systems evolve to a most probable state associated with maximum

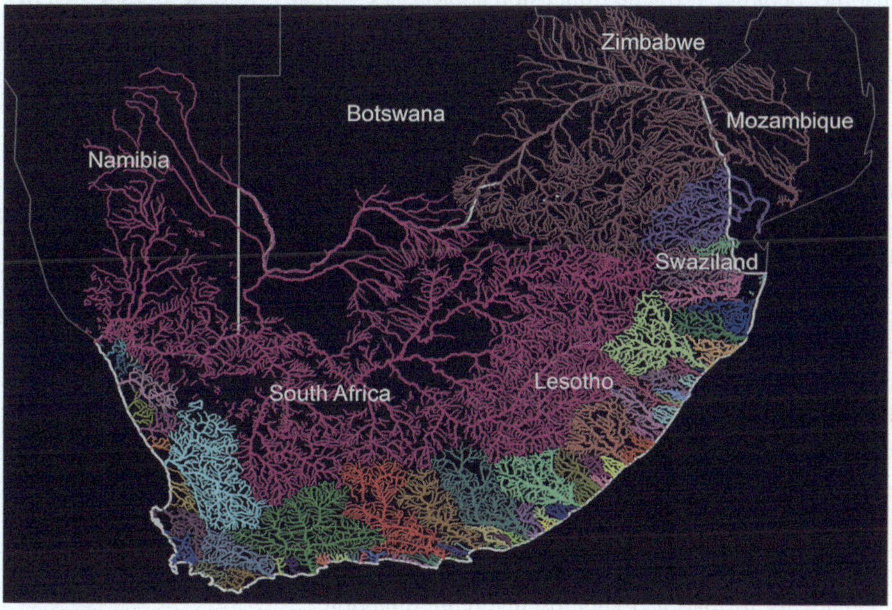

Fig. 6.8 Drainage basins and channel networks in southern Africa. (Reprinted under CC-BY-ND-4.0 license from Mantel & Januchowski-Hartley, 2019. © 2019, The Authors)

entropy. They produced simulated networks with realistic statistical properties based on a random walk. Scheidegger (1967) and Howard (1971) also produced random-branching network models. Woldenberg (1969) argued that the number of stream orders and number of basins per order balances opposing tendencies for minimum overland flow in small basins and maximum work savings in large ones. He was able to make reasonable predictions for 17 real basins based on this model. Kirkby (1971) seems to have been the first to explicitly propose that fluvial system evolution is governed by a principle of optimizing energy dissipation rates, implying that a network pattern that achieves this is likely to be positively selected.

Increasingly sophisticated probabilistic models continued to appear through the early 1980s (Abrahams, 1984). Van Pelt et al. (1989) proposed that channel network topology (and other binary branching patterns) can be treated as a set of partitions of flux magnitudes at junctions. Their models based on branching probabilities gave good fits to 13 of 14 published data sets examined. Combining elements of random topology and spatial simulation models, Troutman and Karlinger (1992, 1994) generated networks on a spatial grid. A Gibbs distribution, with a parameter varying with slope and drainage area was found to be an improvement over a strictly random model.

Several observers in the 1980s noted that channel networks have non-integer fractal dimensions that are often close to 2, the expected value for a space-filling network. This stimulated dozens of studies fitting fractal statistics to maps or digital elevation models of real networks, fractal simulations of networks and basin topography, and arguments that networks evolve to completely drain their catchment, so that the topological fractal dimension approaches 2 (many of these studies are covered in Rodriguez-Iturbe & Rinaldo, 1997).

Simulation models based on principles of self-organization and self-organized criticality in the same time frame were typically closely related to the fractal studies. Some were linked to some controlling geomorphic factor, such as variable substrate erodibility in Stark's (1991) self-avoiding percolation model, and Moglen and Bras's (1995) model based on horizontal and vertical variations in erodibility. Others were simply algorithms that produced plausible channel networks, but with limited hydrologic or geomorphic realism.

Fractal and multifractal structures of stream networks were eventually linked to least-work principles, just as the random models were earlier. Other work explicitly linked the self-organized, fractal structures of channel networks and drainage basins to principles of optimality in terms of energy dissipation (see syntheses in Rodriguez-Iturbe & Rinaldo, 1997; Hergarten, 2013). Other successful (in terms of producing realistic stream networks) models (following Kirkby, 1971) were based on the principle of instabilities in overland flow. Smith and Bretherton (1972) produced a model whereby instabilities leading to channel initiation occur when advective processes dominate diffusive processes. Small disturbances grow rapidly to form channels. Smith (2010) later included these ideas in a more advanced model of channel and network formation, explicitly involving selection.

Branching, tree-like (dendritic) networks typically work best for collecting flow from an area or volume and conveying it to a single point, or for distributing material from a single source to numerous destinations. Patterns that move a flow system toward such topologies are favored by selection, and once formed (even though their nodes and paths may not be permanently fixed) they tend to persist.

Fluvial networks may vary from the dendritic where geological constraints promote other patterns via resistance and gradient selection or prevent the formation of dendritic patterns. Networks of karst conduits, for example, are strongly influenced by geological controls and only occasionally assume a dendritic pattern (Palmer, 1991). Efficiency advantages of branching dendritic systems may conflict with other types of selection, and local selection may not always contribute to overall efficiency of flow systems.

Networks cannot expand indefinitely. Outer, lower-order network links are smaller than higher-order links, and they cannot become indefinitely small. In the case of fluvial networks, a minimum amount of contributing area is required to produce flow sufficient to form a channel or to maintain it by preventing infilling.

Drainage density is channel length per unit area, and its inverse is the contributing area per unit of channel length. If a network is fully developed, the inverse of drainage density of called the constant of channel maintenance. In early stages of network growth, drainage density generally increases rapidly, tapering off over time and reaching a relaxation-time equilibrium where density remains approximately constant (Schumm, 1997).

Climate and geology strongly affect drainage density. Other things being equal, however, if selection favors a most efficient density, then drainage density should increase with the amount of water available. Zhou et al. (2014) found this to be the case in a study of tidal networks, where substrate is relatively uniform and discharge per unit area is determined by tidal range. The direct variation of drainage density with amount of water is also supported by studies of expansion and contraction of wet channels during higher and lower flow conditions (Godsey & Kirchner, 2014). Stream networks may expand or contract in wetter or drier conditions, and those changes vary according to the runoff generating process, Zanetti et al. (2024) showed, using a model incorporating both surface and subsurface flow components, that during saturation-excess runoff generation the stream network expands gradually from the outlet towards the basin divides. Inflow per unit of runoff contributing area is relatively consistent throughout the network, consistent with a linear relationship between discharge and total active stream length. With infiltration-excess runoff, by contrast, more heterogeneous subsurface as well as surface conditions are produced. This is especially pronounced in the wetting phase, where a variable, non-uniform spatial distribution of specific inflow emerges. A hysteretic length vs. discharge relationship and pronounced asymmetry between wetting and drying phases reflects this.

In a classic early study Carlston (1963) studied relationships among drainage density, discharge, and groundwater levels. The key relationship is

$$T = \frac{\left(R / D^2\right)}{8h_o} \tag{6.21}$$

T is transmissivity, R is groundwater recharge, D is drainage density, and h_o the height of the water table at the groundwater drainage divide. Carlston found that, for a constant recharge and h_o, flood discharge varied directly and baseflow inversely and with D^2 in 15 drainage basins of varying lithology and topography, Thus, he concluded, drainage density is adjusted to the most efficient removal of flood runoff. Pallard et al. (2009) reached similar conclusions based on analysis of relationships between flood statistics and drainage density. Combined with the tendency of drainage density to evolve to a steady-state maximum, this shows that there is no advantage to extending channels and networks beyond this point.

The statistical and topological properties of branching dendritic networks can be explained entirely from a mathematical perspective that applies regardless of whether the network comprises stream channels or some other phenomenon (Kovchegov et al., 2022). This provides useful and intriguing avenues for estimating and modeling aspects of fluvial (and other) networks based entirely on combinatorial aspects. However, it does not explain *why* selection prefers branching dendritic networks.

Overall, the research literature shows that realistic stream channel networks can be produced by models with widely varying assumptions, vastly different algorithms, and based on several different underlying principles. Model success in producing realistic outcomes does not necessarily imply that the model is correct, of course. This type of success can be achieved by mimicry where equifinality exists, by skillful tuning and adjustment of model parameters (sometimes called "knob twiddling"), or by cases where inaccurate assumptions are nonetheless useful simplifications for modeling (as in percolation theory). Canonical examples of these phenomena include effective methods for navigation that assume Earth is the center of the solar system, economic models assuming all individuals have complete knowledge and always act to optimize their own profits, and Earth science and ecological models assuming steady-state attractors. This suggests two non-exclusive possibilities: (1) "Model equifinality"(Beven, 1996): where multiple models produce similar outcomes, the channel network is at least semi-independent of the rules under which it is modeled. (2) The varied underlying explanations implied in network models are specific cases of some broader general unifying principle.

The fact that plausible networks can be produced from models with little or no geomorphic or hydrologic basis suggests that model equifinality clearly exists. But does a broader unifying principle also exist?

Branching tree networks work well for performing the hydrological job of collecting and transporting excess moisture. Why? As with the origin of the networks, several different theories and models lead to the same general destination—branching tree-like networks are the most efficient way to get the job done. The most convincing of the numerous and varied explanations and evaluations of this phenomena are based on energetics and are broadly consistent with the least action principle and efficiency selection. For example, a sufficient explanation, Rodriguez-Iturbe et al. (1992) maintained, is based on three criteria: Minimum energy expenditure in any one link; equal energy expenditure per unit bed area throughout the network, and minimum energy expenditure throughout the network. The first and third items follow from the least action principle, so the equal energy expenditure per unit bed area is the only additional assumption that needs to be made. In general, the explanations converge to energy-based maximization of flow efficiency (Nanson & Huang, 2018).

So, the emergence of branching, tree-like networks is favored by selection. But *how* do such patterns emerge, repeatedly, in so many different settings? For this to occur, channels must converge, two at a time, proceeding from the sources to the outlet, or moving upstream from the outlet, channels must bifurcate. This tendency to bifurcation can be explained using a few basic principles of runoff and fluvial hydrology and geomorphology.

1. When concentrated flows intersect, they combine into a single pathway due to the greater efficiency associated with reduced wetted perimeters (and thus reduced friction) and greater mean depth.
2. Intersection of >2 concentrated flows at a single point is unlikely. When it does occur, flow piracy typically prunes all but two channels.
3. As channels extend upslope by headward migration, ground surface slopes are locally increased near the channel and water tables are drawn down. Where anisotropy exists in, e.g., microtopography, surface roughness, and soil properties (as it always does), contributing flows to the extending channel become concentrated. In some cases, the flow concentration is sufficient to initiate side tributary channels by surface incision or pipe erosion (Fig. 6.9).
4. Formation of multiple side tributaries at the same cross-section is unlikely in general, and especially so where asymmetry in the parent channel exists.

Fig. 6.9 Headcut on extending channel near Calvert, Texas. At top, note variations in topography and vegetation along channel, and bank failures which may be sites of side channel formation. Bottom shows closeup of headcut with a bifurcation underway

Non-dendritic channel patterns are nearly always associated with geological controls and constraints, explicable based on resistance and gradient selection.

To explain the advantages of a branching dendritic network, Rodriguez-Iturbe and Rinaldo (1997) used the metaphor of "socialist" vs. "capitalist" approaches to route flow from discrete sources to a single outlet. The socialist network seeks to minimize the total distance for all sources, and routes flow from one to another in sequence to reach the outlet. The capitalist network seeks to minimize distance (and maximize efficiency) for each individual source, resulting in a radiation pattern with each source connected directly to the outlet. A branching dendritic network, they explained, balances the two goals of capitalism and socialism.

As discussed above, multiple theories and explanations lead to this destination, and I indulge myself to add one more, specifically based in network and graph theory. This is based on the advantages of a bifurcating tree graph/network compared to alternatives. The "trunk" of the tree is the basin outlet. Moving upstream, there occurs a series of splits or bifurcations, continuing to the outer, source tributaries. The basin outlet, the source points, and the junctions or bifurcation points represent the nodes of the graph, and the (channel) connections between them are the links. In such a bifurcating tree, $N = 2^q - 1$, where q is the number of stages where bifurcations occur. At the bottom/beginning, $q = 1$, and the outlet point bifurcates into two others. At each q, each point bifurcates into two others. Except for a simple linear sequence or cycle, for $N > 3$ no graph configuration has a lower spectral radius, which is equal to 1.414 no matter how many bifurcations occur. Similarly, algebraic connectivity for a bifurcating graph = 1 for any N. A network such as a channel network can therefore expand indefinitely in this mode without changing the graph complexity or synchronization. Actual drainage networks are often not true bifurcating trees, but the efficiency of this structure probably accounts (consistent with other explanations) for the tendency for selection to promote branching dendritic patterns.

6.5 Rock Fractures

Rock fractures organize as networks, with substantial intra- and inter-fracture network spatial variability (Prabhakaran, Bertotti, et al., 2021a). Valentini et al. (2007) showed that natural rock fracture networks have small world network topological properties associated with high transport efficiency. However, they did not discuss how or why fracture networks with these properties develop. Applying graph theory methods to a dataset including nearly 350,000 ground-truthed fracture traces, Prabhakaran, Urai, et al. (2021b) found that nodes of degree 3 (Y-nodes) were the most common, reflecting fracture growth patterns. This may be another example of selection for branching patterns.

Equivalent fracture network permeability (EFNP) was introduced by Jafari and Babadagli (2013), combining percolation theory and fractal geometry. They represented fracture networks by scanning fracture lines and intersection points to produce a network or lattice where $P(p)$ is the probability that a site is connected to the spanning cluster, and p, p_c are the observed and critical probabilities of site occupation.

$$P(p) \propto (p - p_c)^{\beta}, p > p_c \tag{6.22}$$

The so-called universal exponent β is 0.139 for 2-D and 0.4 for 3-D spaces (and is independent of lattice or continuum percolation). Assuming a relationship between effective permeability and system percolation properties, effective permeability K_{eff} is

$$K_{eff} \propto (\rho - \rho_c)^{\mu} \tag{6.23}$$

Again, μ is a universal exponent.

Percolation theory can be applied in a continuum system by using the excluded volume concept. Jafari and Babadagli (2013) calculated the 2-D excluded area of fractures using

$$A_{ex} = \frac{2}{\pi} l^2 \tag{6.24}$$

where l is the length of fractures in the pattern. Fracture density (fractures per unit area) is

$$\rho = \frac{N_{fr}}{L^2} \tag{6.25}$$

N_{fr} is the number of fractures and L^2 is the domain area. Dimensionless p is then

$$\rho' = \rho A_{ex.} \tag{6.26}$$

Using this estimate, macroscopic properties of fracture networks are independent of fracture shape. They used a constant continuum percolation threshold value of 3.6 in terms of the dimensionless density ρ'. Using a different value would change the quantitative results of their analysis, but not the general relationships identified.

For a given number of fractures, as fracture length increases, the fractal dimension of the scanning lines in the X- and Y-directions decreases, especially near the percolation threshold, where $(p - p_c)$ undergoes continuous increase.

As the fractal dimension increases and approaches two (indicating a space-filling fracture pattern within the domain) the chances of having a percolating network rises, and the dimension density increases accordingly. At higher dimensionless densities the curves converge, representing very well-developed fracture networks. The larger $(p - p_c)$ becomes, due to more or longer fractures, the higher the EFNP value. Thus, by adding a proportionality constant A to Eq. (6.23),

$$EFNP = K_{eff} = A(\rho - \rho_c)^{\mu}. \tag{6.27}$$

Because in most circumstances rock fractures cannot disappear or grow smaller, one interpretation of Jafari and Babadagli's results is that the growth of fracture patterns increases EFNP. However, it is difficult to link this to any specific selection processes or to the "fitness" of the fracture network, especially with respect to purely mechanical fractures produced by stresses on or within rock masses.

However, to the extent fracture networks can be said to grow due to positive feedbacks of weathering, fluid flows within the fractures, and occupation by roots and other biota, there is at least a tentative case to be made for network selection for greater effective permeability.

Fractures or other large-scale permeability structures often serve as preferential flow paths for fluids, but the existence of such structures is not required for PFP formation (Tang et al., 2021). If variations in pore size and/or connectivity are large enough, this can also lead to concentration of flows along preferential pathways. Tang et al. (2021) identified gradual transitions from more-or-less uniform flow fields in more homogeneous, highly connected networks to flow concentration in PFPs with increasing heterogeneity of pore size and decreased connectivity.

With respect to preferential flow through discrete fracture networks in rocks, close relationships exist among preferential flow, network selection, gradient, and resistance selection. A graph theory method based on flow resistance of fracture trace lines was used to identify preferential flow paths by Ma et al. (2024). Fluid tends to flow along the path of least flow resistance, and the minimum resistance and preferential flow paths were found to coincide.

Joints and fractures and subsurface flow networks are inherently difficult to observe and measure, obliging geophysicists, hydrologists, and engineers to rely on modeling. One approach to this conundrum is to develop models which can be constrained or conditioned by whatever observations are available. One example is the KarstNSim model (Gouy et al., 2024), a stochastic model that incorporates however much or little field data is available to simulate discrete karst networks. Unlike models of karst conduit evolution, this network model applies where breakthrough has already occurred (see Ch. 5). It is based on computation of the shortest path between the inlets (e.g., dolines, swallets) and outlets (e.g., springs) of the network using an anisotropic cost function defined on an n-nearest neighbor graph constrained by lithological and structural heterogeneities. The cost function C is an indication of resistance and transport efficiency:

$$C = L\left(a_I C_I + a_F C_F + a_{WT} C_{WT} + a_p C_p\right) \tag{6.28}$$

L is the length of a transport path between nodes (fracture intersections). The individual cost functions account for influence of the distance from inception surfaces (structural faults and horizons) C_I, effects of fractures C_F, (possible multiple) water table influences, accounting for both phreatic and vadose zones C_{WT}, and karstification potential of the rock C_p. The a are weights whose computation is described by Gouy et al. (2024).

Simulated karst networks based on shortest paths defined by C produced realistic results, and conformity with a well-studied karst region in Europe. As noted above, model success in producing realistic results can be achieved for several possible reasons, but the KarstNSim results are at least consistent with network selection for maximum flow efficiency and minimum resistance.

6.6 Scale-Free Networks

Scale-free networks (SFN) are characterized by a power law degree distribution, where most nodes have only a few connections, while a small number of nodes (hubs) have many connections. These networks are self-similar, accounting for the scale-free name, and most form due to some sort of preferential attachment, with high-degree hubs typically attracting more new connections. If these networks are rewired (at least in a way that preserves the degree distribution) their connectivity is preserved. The internet and social networks are canonical examples, along with protein interaction networks in cells.

SFN are robust to random failures, as most nodes are not critical, and allow for efficient information exchange. The preferential attachment property also makes SFN adaptable to changes because new nodes are more likely to connect to highly connected nodes, maintaining system structure and functions. However, SFN are vulnerable to failures of hubs—e.g., attacks in the case of the internet, and major disturbances in the case of ESS.

Scale-free distributions have been found for several geophysical phenomena, including earthquake magnitudes and the size distributions of sinkholes (Steacy & McCloskey, 1999; Abe & Suzuki, 2006; Yizhaq et al., 2017). Power law distributions are ubiquitous in Earth phenomena, and self-similarity over two or more orders of magnitude of spatial and temporal scale are as well (e.g., Hergarten, 2013; Kovchegov et al., 2022). However, power-law distributions are just one class of heavy-tailed distributions that are often fit as well or better by other functions, and self-similarity over several orders of magnitude may or may not be relevant to geophysical phenomena that may vary over many orders of magnitude. Finally, power-law distributions of phenomena such as earthquake sizes or landslides (while very relevant in other contexts), are not germane to the topic of selection of flux or interaction networks in ESS.

One of the few studies of ESS flux networks in this context is that of Castillo et al. (2020), who analyzed patterns of surface water connectivity in a lowland river/floodplain system (Mission River, Texas). They presented evidence that connectivity patterns typically show scale-free network properties consistent with optimization of energy expenditure. This suggests that in other cases scale-free patterns may be selected for, consistent with thermodynamic selection for optimal energy expenditure.

References

Abe, S., & Suzuki, N. (2006). Complex-network description of seismicity. *Nonlinear Processes in Geophysics, 13*, 145–150. https://doi.org/10.5194/npg-13-145-2006

Abrahams, A. D. (1984). Channel networks: a geomorphological perspective. *Water Resour. Res. 20*, 161e188.

Berkowitz, B., & Balberg, I. (1993). Percolation theory and its application to groundwater hydrology. *Water Resources Research, 29*, 775–794.

Berkowitz, B., & Ewing, R. P. (1998). Percolation theory and network modeling applications in soil physics. *Surveys in Geophysics, 19*, 23–72.
Beven, K. (1996). Equifinality and uncertainty in geomorphological modelling. In B. Rhoads & C. Thorn (Eds.), *The scientific nature of geomorphology, proceedings of the 27th Binghamton Geomorphology Symposium* (pp. 289–313). John Wiley.
Carlston, C. W. (1963). Drainage density and streamflow. *U.S. Geological Survey Professional Paper, 422-C*, C1–C8.
Castillo, C. R., Güneralp, I., Hales, B., & Güneralp, B. (2020). Scale-free structure of surface water connectivity within a lowland river floodplain landscape. *Geophysical Research Letters, 47*, e2020GL088378.
Cong, Z., Li, Q., Mo, K., et al. (2017). Ecohydrological optimality in the Northeast China Transect. *Hydrology and Earth System Sciences, 21*, 2449–2462.
Doolittle, W. F. (2017). Darwinizing Gaia. *Journal of Theoretical Biology, 434*, 11–19.
Egli, M., Hunt, A. G., Dahms, D., et al. (2018). Prediction of soil formation as a function of age using the percolation theory approach. *Frontiers in Environmental Science, 6*, 108.
Geller, W., Kitchens, B., Misiurewicz, M., & Rams, M. (2012). A spectral radius estimate and entropy of hypercubes. *International Journal of Bifurcation and Chaos, 22*, 1250096. https://doi.org/10.1142/S0218127412500964
Ghanbarian, B. (2021). Unsaturated hydraulic conductivity in dual-porosity soils: Percolation theory. *Soil and Tillage Research, 212*, 105061.
Godsey, S. E., & Kirchner, J. W. (2014). Dynamic, discontinuous stream networks: Hydrologically driven variations in active drainage density, flowing channels and stream order. *Hydrological Processes, 28*, 5791–5803.
Goodwell, A. E., Kumar, P., Fellows, A. W., & Flerchinger, G. N. (2018). Dynamic process connectivity explains ecohydrologic responses to rainfall pulses and drought. *Proceedings of the National Academy of Sciences (USA), 115*, E8604–E8613. https://doi.org/10.1073/pnas.1800236115
Goodwell, A. E., Jiang, P., Ruddell, B. L., & Kumar, P. (2020). Debates—Does information theory provide a new paradigm for Earth science? Causality, interaction, and feedback. *Water Resources Research, 56*, e2019WR024940.
Gouy, A., Collon, P., Bailly-Compte, V., et al. (2024). KarstNSim: A graph-based method for 3D geologically-driven simulation of karst networks. *Journal of Hydrology, 632*, 130878.
Hembry, D. H., Raimundo, R. L. G., Newman, E. A., et al. (2018). Does biological intimacy shape ecological network structure? A test using a brood pollination mutualism on continental and oceanic islands. *Journal of Animal Ecology, 87*, 1160–1171.
Hergarten, S. (2013). *Self-organized criticality in earth systems* (2nd ed.). Springer.
Howard, A. D. (1971). Simulation of stream networks by headward growth and branching. *Geographical Analysis, 3*, 29–50.
Huang, H. Q., & Nanson, G. C. (2000). Hydraulic geometry and maximum flow efficiency as products of the principle of least action. *Earth Surface Processes and Landforms, 25*, 1–16.
Huang, H. Q., & Nanson, G. C. (2007). Why some rivers develop an anabranching pattern. *Water Resources Research, 43*, W0y441. https://doi.org/10.1029/2006WR005223
Hunt, A. G. (1998). Upscaling in subsurface transport using cluster statistics of percolation. *Transport in Porous Media, 30*, 177–198.
Hunt, A. G. (2016). Spatio-temporal scaling of vegetation growth and soil formation from percolation theory. *Vadose Zone Journal, 15*, 1–15. https://doi.org/10.2136/vzj2015.01.0013
Hunt, A. G. (2017). Use of constructal theory in modeling in the geosciences. In B. Ghanbarian & A. G. Hunt (Eds.), *Fractals. Concepts and Applications in Geosciences*. CRC Press, Ch. 12, 15 p.
Hunt, A. G. (2024). The physics and the biology of the water balance: A personal journey through the critical zone into the water balance. *Hydrological Processes, 38*, e15209.
Hunt, A. G., & Ghanbarian, B. (2016). Percolation theory for solute transport in porous media: geochemistry, geomorphology, and carbon cycling. *Water Resources Research, 52*, 7444–7459. https://doi.org/10.1002/2016WR019289

Hunt, A. G., Ghabarian, B., Faybishenko, B. (2023). A model of temporal and spatial river network evolution with climatic inputs. *Frontiers in Water 5*, 1174570.

Hunt, A. G., Egli, M., & Faybishenko, B. A. (2021). *Hydrogeology, chemical weathering, and soil formation, geophysical monographs*. AGU/Wiley.

Hunt, A. G., Sahimi, M., Ghanbarian, B., & Poveda, G. (2024). Predicting ecosystem net primary productivity by percolation theory and optimality principle. *Water Resources Research, 60*, e2023WR036340.

Jafari, A., & Babadagli, T. (2013). Relationship between percolation–fractal properties and permeability of 2-D fracture networks. *International Journal of Rock Mechanics and Mining Sciences, 60*, 353–362.

Jia, Y. (1990). Minimum Froude number and the equilibrium of alluvial sand rivers. *Earth Surface Processes and Landforms, 15*, 199–208.

Kirkby, M. J. (1971). Hillslope process-response models based on the continuity equation. *Institute of British Geographers Special Publication, 3*, 15–30.

Kovchegov, Y., Zaliapin, I., & Foufoula-Georgiou, E. (2022). Random self-similar trees: Emergence of scaling laws. *Surveys in Geophysics, 43*, 353–421.

Lapenis, A. G. (2002). Directed evolution of the biosphere: Biogeochemical selection or Gaia? *Professional Geographer, 54*, 379–391.

Lekevicius, E. (2002). *The origin of ecosystems by means of natural selection*. Lithuanian Academy of Sciences, Institute of Ecology, Vilnius.

Lekevicius, E. (2006). The Russian Paradigm in ecology and evolutionary biology: Pro et contra. *Acta Zoologica Lithuania, 16*, 3–19.

Leopold, L. B., & Langbein, W. B. (1962). *The concept of entropy in landscape evolution*. US Geological Survey Professional Paper 5OOA.

Li, G., & Good, S. P. (2021). Information-based uncertainty decomposition in dual-channel microwave remote sensing of soil moisture. *Hydrology and Earth System Sciences, 25*, 5029–5045.

Lotka, A. J. (1922). Contribution to the energetics of evolution. *Proceedings of the National Academy of Sciences, 8*, 147–151.

Ma, L., Cui, X. L., Zhang, C. C., et al. (2024). Effective method for identification of preferential flow paths in two-dimension discrete fracture methods based on a flow resistance method. *Hydrogeology Journal, 32*, 967–982.

Mantel, S., & Januchowski-Hartley, S. (2019, February 3). What we learned when our map of southern Africa's rivers went viral. *The Conversation*. https://theconversation.com/what-we-learned-when-our-map-of-southern-africas-rivers-went-viral-110735 (last accessed 14 September 2024).

Moglen, G. E., & Bras, R. L. (1995). The effect of spatial heterogeneities on geomorphic expression in a model of basin evolution. *Water Resources Research, 30*, 2489–2498.

Mowshowitz, A., & Dehmer, M. (2012). Entropy and the complexity of graphs revisited. *Entropy, 14*, 559–570.

Nanson, G. C., & Huang, H. Q. (2008). Least action principle, equilibrium states, iterative adjustment and the stability of alluvial channels. *Earth Surface Processes and Landforms, 33*, 923–942.

Nanson, G. C., & Huang, H. Q. (2017). Self-adjustment in rivers: Evidence for least action as the primary control of alluvial-channel form and process. *Earth Surface Processes and Landforms, 42*, 575–594.

Nanson, G. C., & Huang, H. Q. (2018). A philosophy of rivers: Equilibrium states, channel evolution, teleomatic change and least action principle. *Geomorphology, 302*, 3–19.

Paik, K., & Kumar, P. (2010). Optimality approaches to describe characteristic fluvial patterns on landscapes. *Philosophical Transactions of the Royal Society B, 365*, 1387–1395.

Pallard, B., Castellarin, A., & Montanari, A. (2009). A look at the links between drainage density and flood statistics. *Hydrology and Earth System Sciences, 13*, 1019–1029.

Palmer, A. N. (1991). Origin and morphology of limestone caves. *Geological Society of America Bulletin, 103*, 1–21.

Pelletier, J. D. (2008). *Quantitative modeling of earth surface processes*. Cambridge University Press.

Phan, H. C., Dhar, A. S., Thodi, P., et al. (2018). Probability of network disconnection of water distribution system for maintenance prioritization. *Journal of Water Supply Research and Technology, 67*, 252–269.

Phan, H. C., Dhar, A. S., & Bui, N. D. (2021). Accounting for source location on the vulnerability assessment of water distribution network. *Journal of Infrastructure Systems, 27*, 04021024.

Phillips, J. D. (1990). The instability of hydraulic geometry. *Water Resources Research, 26*, 739–744.

Phillips, J. D. (1991). Multiple modes of adjustment in unstable river channel cross- sections. *Journal of Hydrology, 123*, 39–49.

Phillips, J. D. (2010). The job of the river. *Earth Surface Processes and Landforms, 35*, 305–313.

Phillips, J. D. (2011). Emergence and pseudo-equilibrium in geomorphology. *Geomorphology, 132*, 319–326.

Phillips, J. D. (2013). Networks of historical contingency in Earth surface systems. *Journal of Geology, 121*, 1–16.

Phillips, J. D. (2016a). Complexity of Earth surface system evolutionary pathways. *Mathematical Geosciences, 48*, 743–765.

Phillips, J. D. (2016b). Vanishing point: Scale independence in geomorphic hierarchies. *Geomorphology, 266*, 66–74.

Phillips, J. D. (2022a). Geomorphic impacts of Hurricane Florence on the lower Neuse River: Portents and particulars. *Geomorphology, 397*, 108026.

Phillips, J. D. (2022b). Geomorphology of the fluvial-estuarine transition zone, Neuse River, North Carolina. *Earth Surface Processes and Landforms, 47*, 2044–2061.

Phillips, J. D. (2023). Why everything is connected to everything else. *Ecological Complexity, 54-55*, 101051.

Prabhakaran, R., Bertotti, G., Urai, J., & Smeulders, D. (2021a). Investigating spatial heterogeneity within fracture networks using hierarchical clustering and graph distance metrics. *Solid Earth, 12*, 2159–2209.

Prabhakaran, R., Urai, J. L., Bertotti, G., et al. (2021b). Large-scale natural fracture network patterns: Insights from automated mapping in the Lilstock (Bristol Channel) limestone outcrops. *Journal of Structural Geology, 150*, 10445.

Restrepo, J. G., Ott, E., & Hunt, B. R. (2006). Emergence of synchronization in complex networks of interacting dynamical systems. *Physica D, 224*, 114–122.

Rodriguez-Iturbe, I., & Rinaldo, A. (1997). *Fractal River Basins. Chance and self-organization.* Cambridge University Press.

Rodriguez-Iturbe, I., Rinaldo, A., Rigon, R., et al. (1992). Energy dissipation, runoff production, and the three-dimensional structure of river basins. *Water Resources Research, 28*, 1095–1103.

Scheidegger, A. E. (1967). A stochastic model for drainage patterns into an intermontane trench. *Hydrological Sciences Bulletin, 12*, 15–20.

Schumm, S. A. (1997). Drainage density. Problems of prediction and application. In D. R. Stoddart (Ed.), *Process and form in geomorphology* (pp. 15–45). Routledge.

Sheldrake, M. 2020. Entangled Life: How fungi make our worlds, change our minds and shape our futures. , .

Smith, C. H. (1986). A contribution to the geographical interpretation of biological change. *Acta Biotheoretica, 35*, 229–278.

Smith, T. R. (2010). A theory for the emergence of channelized drainage. *Journal of Geophysical Research: Earth Surface, 115F*, F02023.

Smith, T. R., & Bretherton, F. P. (1972). Stability and the conservation of mass in drainage basin evolution. *Water Resources Research, 8*, 1506–1529.

Stark, C. P. (1991). An invasion percolation model of drainage network evolution. *Nature, 352*, 423–425.

Steacy, S. J., & McCloskey, J. (1999). Heterogeneity and the earthquake magnitude-frequency distribution. *Geophysical Research Letters, 26*, 899–902.

Strydom, T., Dalla Riva, G. V., & Poisot, T. (2021). SVD entropy reveals the high complexity of ecological networks. *Frontiers in Ecology and Evolution, 9*, 623141.

Summerfield, M. A. (1991). *Global geomorphology*. Longman.

Tang, Y. B., Zhao, J. Z., Bernabé, Y., & Li, M. (2021). Fluid flow concentration on preferential paths in heterogeneous porous media: Application of graph theory. *Journal of Geophysical Research: Solid Earth, 126*, e2021JB023164.

Tennant, C., Larsen, L., Bellugi, D., et al. (2020). The utility of information flow in formulating discharge forecast models: A case study from an arid snow-dominated catchment. *Water Resources Research, 56*, e2019WR024908.

Tiwari, S., Recinos Brizuela, S., Hein, T., et al. (2024). Water-controlled ecosystems as complex networks: Evaluation of network-based approaches to quantify patterns of connectivity. *Ecohydrology, 17*, e2690. https://doi.org/10.1002/eco.2690

Troutman, B. M., & Karlinger, M. R. (1992). Gibbs' distribution on drainage networks. *Water Resources Research, 28*, 563–577.

Troutman, B. M., & Karlinger, M. R. (1994). Inference for a generalized Gibbsian distribution on channel networks. *Water Resources Research, 30*, 2325–2338.

Valentini, L., Perugini, D., & Poli, G. (2007). The "small-world" topology of rock fracture networks. *Physica A, 377*, 323–328.

Van Nuland, M. E., Wooliver, R. C., Pfennigwerth, A., et al. (2016). Plant-soil feedbacks: Connecting ecosystem ecology and evolution. *Functional Ecology, 30*, 1032–1042.

Van Pelt, J., Woldenberg, M. J., & Verwer, R. W. H. (1989). Two generalized topological models of stream network growth. *Journal of Geology, 97*, 281–299.

Verboom, W. H., & Pate, J. S. (2013). Exploring the biological dimension to pedogenesis with emphasis on the ecosystems, soils and landscapes of southwestern Australia. *Geoderma, 211*(212), 154–183.

Ware, I. M., Fitzpatrick, C. R., Senthilnathan, A., et al. (2019). Feedbacks link ecosystem ecology and evolution across spatial and temporal scales: Empirical evidence and future directions. *Functional Ecology, 33*, 31–42.

Woldenberg, M. J. (1969). Spatial order in fluvial systems: Horton's laws derived from mixed hexagonal hierarchies of drainage basin areas. *Geological Society of America Bulletin, 80*, 97–112.

Yazdani, A., & Jeffrey, P. (2012). Applying network theory to quantify the redundancy and structural robustness of water distribution systems. *Journal of Water Resources Planning and Management, 138*, 153–161.

Yizhaq, H., Ish-Shalom, C., Raz, E., & Ashkenazy, Y. (2017). Scale-free distribution of Dead Sea sinholes—Observations and modeling. *Geophysical Research Letters, 44*, 4944–4952.

Zanetti, F., Camporese, M., & Botter, G. (2024). How do different runoff generation mechanisms drive stream network dynamics? Insights from physics-based modelling. *Hydrological Processes, 38*, e15234.

Zhang, Y., Zhao, T., Shi, C., & Ma, Q. (2021). Simulation of vegetation cover based on the theory of ecohydrological optimality in the Yongding River watershed, China. *Forests, 12*, 1377.

Zhou, Z., Stefanon, L., Olabarrieta, M., et al. (2014). Analysis of the drainage density of experimental and modelled tidal networks. *Earth Surface Dynamics, 2*, 105–116.

Thermodynamic Selection

7

Abstract

The laws of thermodynamics, especially the second law, underpin the least action, maximum power, and maximum entropy production principles. They boil down to depleting energy gradients as rapidly and efficiently as possible. As varied forms and pathways of energy flows and transformations "explore" their possibility space, those that achieve the gradient depletion and efficiency criteria are retained in the short term and recur repeatedly over time. The explorations that do not meet these criteria tend to fade and vanish and do not show a pattern of recurrence. These principles can be grouped together as forms of thermodynamic selection. Thermodynamic selection occurs in general processes such as turbulent fluid flows and atmospheric energy exchanges. These are manifested in global phenomena such as atmospheric and ocean general circulation, evaporation and the hydrologic cycle, and teleconnection patterns such as ENSO. Thermodynamic selection is also evident at more local scales, e.g., in fluvial and estuarine processes.

7.1 Thermodynamics and Evolution

Energy and mass are always conserved, in some form, according to the first and second laws of thermodynamics. The energy which is not used to do work must be dissipated as entropy. Work is linked to power, the rate of work. Earth surface systems (ESS) are all governed by thermodynamics, and selection favors certain aspects of energy conversions.

Connections between thermodynamics and natural selection were recognized more than a century ago, most notably by Lotka (1922): "The first effect of natural selection thus operating on competing species will be to give relative preponderance (in number or mass) to those most efficient in guiding available

J. D. Phillips, *Abiotic Selection in Earth Surface Systems*, Geophysics and Environmental Physics, https://doi.org/10.1007/978-3-031-85862-8_7

energy …. Primarily the *path* of the energy flux through the system will be affected (p. 147).” Lotka pointed out that if excess energy inputs are available, the total energy flux can be increased by evolution (or immigration or proliferation) of organisms suited to using it. Where energy is limited, available power can be increased by higher turnover rates. If energy and mass resources are not limiting, selection operates to increase total mass and rates of circulation. Thus Lotka’s (1922) perspective is consistent with the more recent notions of biogeochemical selection.

Under the “appropriate conditions,” Odum and Pinkerton (1955) wrote, maximum power output is the criterion for survival of systems. Thus, survival of the fittest implies that those forms which can produce the greatest power output persist. They defined efficiency in terms of the ratio of mass inputs to outputs and proposed that systems perform at an optimum efficiency for maximum power output, which is less than maximum input/output efficiency. Though Odum and Pinkerton were primarily concerned with ecological systems, they applied their argument to systems in general, including purely mechanical ones.

The self-organization of complex structures is thermodynamically favored if they dissipate free energy that otherwise could not be accessed (Ueltzhöffer et al., 2021). This applies to relatively simple geophysical dissipative systems such as convective cells, lightning strikes, hurricanes, and mechanical cracks as well as to complex biological systems and ESS. Work-harvesting processes are necessary to maintain dissipative structures, and Ueltzhöffer et al. (2021) demonstrated this in a stochastic chemical reaction network. Such structures can emerge naturally in a hierarchy of self-similar dissipative structures, where each level feeds on the dissipation of a lower level, moving away from the initial driving energy gradient.

A common theme is a need or tendency to deplete energy gradients as rapidly as possible (Annila, 2021). In a closed system this leads to eventual equalization of potentials and reduction of gradients. In geophysical and ecological systems, however, gradients are consistently renewed by new energy inputs from solar radiation and radioactive decay within the planet. Selection favors the fastest reduction of energy differentials, Annila and Salthe (2010) maintained, a principle derived from the second law of thermodynamics and which subsumes natural selection. Skene (2015) developed a thermodynamic theory of biological evolution, based on the first and second laws of thermodynamics, diffusion, and the maximum entropy production principle. Because ecosystems and other open systems move towards maximizing dispersal of energy, biological diversity is expected to increase towards a level D_{max} representing maximum entropy production (S_{max}). Skene developed a mathematical model to predict diversity over the last 500 million years, combining diversification, post-extinction recovery and likelihood of discovery of the fossil record, and compared predictions with the fossil record. The model predicts that life diffuses into available energetic space (ecospace) towards a dynamic equilibrium, driven by increasing entropy within the genetic material. This dynamic equilibrium is punctured by extinction events, which are followed by restoration of D_{max} through diffusion into available ecospace. Skene's (2015) ecospace is comparable to the resource space of Phillips (2008) and the multidimensional niche concept of Hutchinson (1957). The model is also consistent with idea of biological saturation. Cushman (2023) also supports the concept of evolution based on dissipation of free energy.

Energy dissipation—specifically, photon dissipation of solar radiation in the UV-C wavelengths—is the driving factor of evolution according to Michaelian (2011, 2023). Michaelian describes a stochastic "natural thermodynamic selection" of characteristics of structures (or processes) that increase the global rate of dissipation of the prevailing solar spectrum. At different biotic-abiotic levels, different mechanisms of thermodynamic selection are prominent, from the molecular level at the origin of life, up to the level of the present biosphere with non-linear coupling of biotic and abiotic processes. At the levels of the organism and the biosphere, respectively, the non-equilibrium thermodynamic description of evolution is broadly consistent with the Darwinian and Gaian descriptions, even though the underlying mechanisms and the functions of selection differ.

7.2 Atmospheric Circulation

As an illustration of thermodynamic selection at work, we turn to global patterns of tropics-to-poles heat transport and general atmospheric circulation.

Consider the generation of free energy A from solar radiation. This represents work or power using a low-entropy energy source. Resulting free energy can be converted further into other forms, and it can be dissipated. A free energy budget is reflected by

$$\frac{dA}{dt} = G - D - G_{conv}, \tag{7.1}$$

where G is free energy generation, G_{conv} is conversion to other forms, and D is dissipation (entropy). The global climate and circulation system can be considered as a heat engine (Kleidon, 2023). Using a power plant analogy, the plant generates heat by burning the free energy stored in its fuel at a rate $J_{in,}$ with some of that turned into electricity at a rate G. Waste heat flux is $J_{out.}$ The first law requires that (with no change in heat content and in steady state), $J_{in} = G + J_{out}$. Also limiting energy conversion is the second law. Heat added to the system comes at a temperature T_{in}, and entropy of the system is increased at a rate J_{in}/T_{in}. Waste heat goes out at a different temperature T_{out}, reducing entropy of the system at a rate of J_{out}/T_{out}. Again assuming steady state and no change in heat storage within the system, the entropy budget is balanced by entropy production within the system (σ):

$$\frac{J_{in}}{T_{in}} + \sigma = \frac{J_{out}}{T_{out}} \tag{7.2}$$

As it generates free energy, G is not part of the entropy budget. By the second law $\sigma \geq 0$. This leads us to the Carnot limit

$$G \leq J_{in}\left(\frac{T_{in} - T_{out}}{T_{in}}\right) \tag{7.3}$$

In the Earth system the energy balance is set by the total absorbed solar radiation $R_{s,tot}$ and the emission of terrestrial radiation to space:

$$R_{s,tot} = \sigma T_r^4 \tag{7.4}$$

T_r is the radiative temperature. Power derived from the heat flux has a maximum power limit achieved with an intermediate optimum value

$$J_{opt} = 0.5\left(R_s + R_{l,down} - R_{l,0}\right). \tag{7.5}$$

The subscripts s, l indicate incoming shortwave and outgoing longwave radiation. $R_{l,0}$ is a constant term originating from linearizing the Stefan-Boltzmann law, $R_{l,0} = \sigma T_o^4$ with T_o a fixed reference temperature. The surface temperature associated with the maximum power is

$$T_s = T_r + \frac{R_s + R_{l,down} - R_{l,0}}{2k_r}. \tag{7.6}$$

The constant $k_r = 4\sigma T_o^3$.

As Kleidon (2023; the basis for the discussion above) puts it, power maximization constrains system dynamics so that the magnitude of the heat flux J and the temperature T_s can be predicted based on the maximization. Only the radiative forcing and some assumptions about how radiative transfer is formulated are then required.

Three forms of entropy are involved in energy conversions in the Earth system—radiative, molar, and thermal, at the levels of photons, electrons, and molecules. The solar energy source in the form of electromagnetic radiation at short wavelengths has low radiative entropy because solar radiation is emitted at a very high temperature. When insolation is absorbed and turned into heat the thermal energy has associated thermal entropy. When sunlight is used by photosynthesis or photovoltaics, some of the energy is converted into non-thermal forms of energy and molar entropy, but not heat.

Following conversion, energy is eventually returned to space as emission of much longer wavelength terrestrial radiation, with much greater radiative entropy (due to emission at far colder temperatures).

Three processes are mainly responsible for generation of free energy from solar radiation. Free energy as kinetic energy in atmospheric motion is generated by heat engines that operate as the power plant example used to illustrate the Carnot limit (Eq. 7.3). Chemical free energy embedded in carbohydrates and oxygen is produced by photosynthesis. These forms of free energy are associated with dynamics such as metabolism and atmospheric motion and can be transformed into other forms of free energy, associated with, e.g., ocean dynamics, hydrological processes, biogeochemical cycling, and ecosystem production.

Thermodynamic optimality and limits of these three processes that transform solar radiation into free energy constrains the dissipative dynamics of the subsequent energy conversions. If Earth system processes are constrained by thermodynamic limits, and if this is associated with optimal functioning that maximizes the free energy generation for Earth system dynamics, these limits should be reflected in observed climatological patterns.

7.2.1 Poleward Heat Transport and General Circulation

Using a realistic but highly simplified model, Kleidon (2023) applied the principles above to Earth's global patterns of heat transport and general circulation. Earth experiences major imbalances in solar input between tropical and poleward regions. This imbalance in solar radiative heating is associated with large-scale motions that enable the atmosphere to generate free energy in the form of kinetic energy. Heat is transported by this motion, thereby influencing the thermodynamic forcing. This combination leads to a maximum power limit. There is therefore a process that performs work by converting a heat difference into free energy, producing motion which is dissipated back into heat (due to friction), and produces thermal entropy.

Kleidon (2021, 2023) demonstrates the maximum power limit using a two-box model, dividing the planet into regions between 30° north and south latitudes, and portions poleward of 30° (Fig. 7.1). Lower-latitude regions absorb more solar radiation per unit surface area than the higher-latitude regions. The imbalance leads to a difference in surface temperatures between the tropics and the extratropics, setting up a heat engine. The work performed by this heat engine generates large-scale atmospheric motion, its associated kinetic energy, and is subject to the Carnot limit.

Surface temperatures are functions of the absorbed insolation and the poleward heat flux J:

$$T_{s,t} = \frac{\left(R_{s,t} - a - J\right)}{b} \tag{7.7}$$

$$T_{s,p} = \frac{\left(R_{s,p} + a + J\right)}{b}. \tag{7.8}$$

The t, p subscripts indicate the low- and high-latitude boxes, respectively. Empirical constants $a = -388.7$ W m^{-2} and $b = 2.17$ W m^{-2} K^{-1} (T is in units of °K) are derived from Budyko (1969).

With no heat transport, $J = 0$ and temperatures are determined solely by local solar radiative forcing. This yields the greatest temperature difference and the

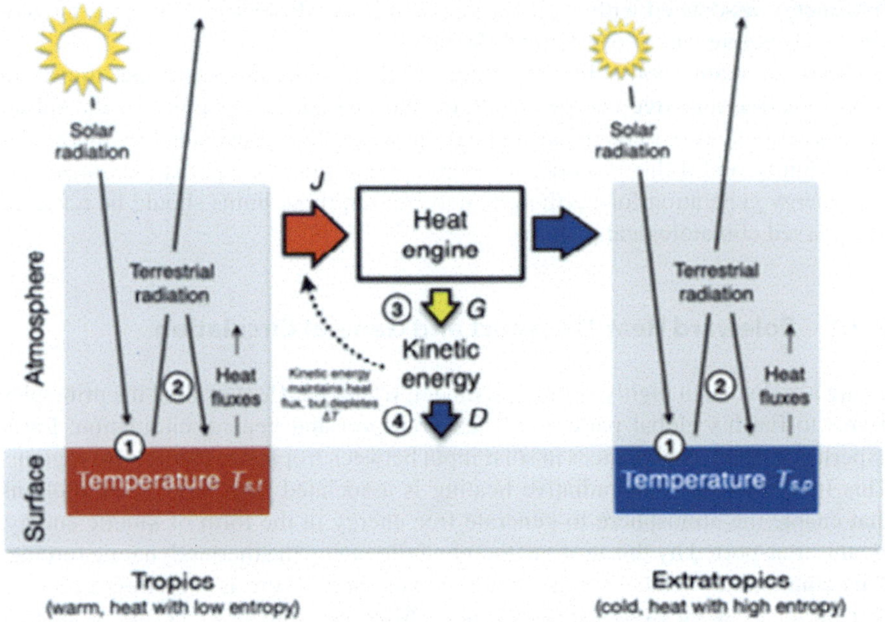

Fig. 7.1 Basis of Kleidon's (2023) two-box model of how the atmospheric heat engine generates kinetic energy, using the radiative heating differences between the tropics and extratropics. The generated motion maintains the poleward heat transport J but depletes the temperature difference $T_{s,t} - T_{s,p}$ between the two boxes. This results in a maximum power limit for the strength of the atmospheric circulation. Numbered circles indicate thermodynamic conversions and entropy production: (1) conversion of solar radiation to thermal energy by absorption at the surface temperature, producing thermal entropy; (2) absorption of terrestrial radiation and reemission (and vertical convection), which both produce thermal entropy; (3) generation of free energy from heat flux J; and (4) frictional dissipation, which converts kinetic energy back into thermal energy, which produces thermal entropy. (Reprinted under CC-BY-4.0 license from Kleidon, 2023. © 2023, The Author)

highest efficiency in the Carnot limit—but no power. With a value of $J_{max} = (R_{s,t} - R_{s,p})/2$, the temperature difference disappears, giving a globally uniform surface temperature of 290°K, and the efficiency in the Carnot limit goes to zero, also yielding no power. A maximum in the power thus exists at an intermediate value of J. Kleidon (2023) calculated a maximum power of

$$G_{max} = \left(R_{s,t} - R_{s,p}\right)^2 / \left(16bT_{s,t}\right). \tag{7.9}$$

Estimates based on maximum power and the Carnot limit are broadly supported by observational data from the Clouds and Earth's Radiant Energy System (CERES) satellite-derived data set (Fig. 7.2). The broad global patterns of heat flux and atmospheric circulation thus reflect thermodynamic selection in the form of maximum power generation, or as Kleidon (2023) put it, "heat engines working as hard as they can."

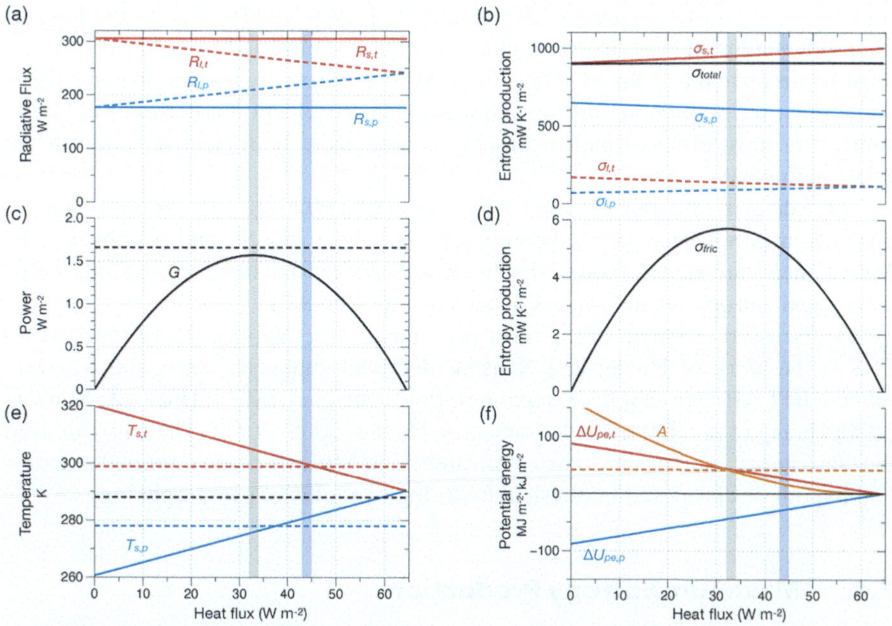

Fig. 7.2 Radiative forcing and energy properties estimated from Kleidon's (2023) two-box model. (**a**) Absorbed solar radiation ($R_{s,t}$ and $R_{s,p}$) of the tropical (red lines) and extratropical box (blue lines) and emitted terrestrial radiation to space ($R_{l,t}$ and $R_{l,p}$). (**b**) Entropy production by absorption of solar radiation ($\sigma_{s,l}$ and $\sigma_{s,p}$), radiative transfer of terrestrial radiation ($\sigma_{l,t}$ and $\sigma_{l,p}$) and total entropy production, σ_{total} (black line). (**c**) Power G to generate atmospheric motion. The horizontal dashed line marks the power inferred directly from the CERES forcing. (**d**) Entropy production by frictional dissipation of kinetic energy. (**e**) Surface temperatures of the tropical (red line) and extratropical (blue line) boxes with the surface temperatures estimated from the CERES top-of-atmosphere flux marked by the dashed lines (black line: global average). (**f**) Difference in potential energy compared to an isothermal Earth ($\Delta U_{pe,t}$, $\Delta U_{pe,p}$, in MJm^{-2}, red and blue lines) and the available potential energy (A, in kJm^{-2}, orange line, dashed orange line represents published estimates). The vertical shaded areas represent the value of heat transport from maximum power (grey) and inferred from the CERES data (blue). (Reprinted under CC-BY-4.0 license from Kleidon, 2023. © 2023, The Author)

7.3 Least Action and Maximum Efficiency

Recall that the least action principle (LAP), described in Chap. 2, states that in a conservative dynamical system the motion between any two points is such that the action has a minimum value with respect to all paths between the points that correspond to the same energy. This is commonly generalized to something along the lines of "nature always finds the most efficient path." In ESS the LAP implies that work, be it fluid flows, sediment transport, heat flux, or biological productivity, is accomplished with as little energy as possible. It does NOT suggest that work itself is minimized (work or power is in fact often maximized), but that the work that is done is accomplished using the minimum possible amount of the available energy.

This is equivalent to a principle of maximum efficiency. Also, due to the laws of thermodynamics, if less of the available energy is converted to work, then more must be dissipated as entropy. Thus, the LAP is consistent with, and in some cases tantamount to, maximum entropy production. As a selection criterion, the LAP's shortest or most efficient path is broadly analogous to biological selection of the fittest individuals.

There do exist situations where LAP ≈ maximum efficiency ≈ maximum entropy production is not necessarily theoretically valid, but in ESS and geophysics the equivalencies are sound, though the target states of least action and maximum efficiency and entropy are not always achieved (see Sect. 8.6).

As discussed in Chap. 2, the most prominent direct application of the LAP to ESS is the work of Huang and Nanson of fluvial channels, where they directly related the LAP to principle of maximum flow efficiency (MFE; Huang & Nanson, 2000; Huang et al., 2004, 2014; Nanson & Huang, 2008, 2017). They also showed the consistency, and in some cases equivalency, of MFE with other optimal or extremal hypotheses proposed in physical hydrology and fluvial geomorphology.

7.4 Maximum Entropy Production

The MEP (maximum entropy production) principle states that systems of sufficient complexity are characterized by non-equilibrium thermodynamic states in which the rate of thermodynamic entropy production is maximized. Applications of MEP in physics, chemistry, and biology are reviewed by Martyushev and Selenev (2006), and Ozawa et al. (2003) surveyed the use of MEP in studies of global climate. Dyke and Kleidon (2010) discussed MEP applications of the Earth system.

MEP is not a physical law, though it is closely linked to the second law of thermodynamics. Dewar (2009) characterized it as an inference algorithm that translates physical assumptions into macroscopic predictions. Dyke and Kleidon (2010) also characterized MEP as an inferential tool that produces the best prediction from the available information. If the physical assumptions underlying MEP in a particular case are invalid, then the predictions will be erroneous. Indeed, like statistical mechanics more broadly, MEP is probabilistic rather than deterministic.

Boltzmann showed that the entropy of a system at thermodynamic equilibrium can be computed by.

$$S = k_B \ln \Omega \tag{7.10}$$

where the Boltzmann constant k_B translates microscopic energy of particles to the macroscopic property of temperature, and Ω is the number of different microstates possible for a given microstate. With large systems Ω is big enough so that statistical characterizations are appropriate.

For Earth's atmosphere and other large systems, the rate of entropy change is

$$\frac{dS}{dt} = \sigma + \sum NEE_i \tag{7.11}$$

where *NEE* is the net entropy exchanged across the boundary of the system and σ the entropy produced within the system. Where $dS/dt = 0$ at steady state,

$$\sigma = F\nabla\left(\frac{1}{T}\right) = -\sum_{i} NEE_i \quad (7.12)$$

where F is heat flux and T is temperature. For the case of Earth's heat balance, if F is the net flow of heat from the tropics poleward, increased internal entropy production is related to more rapid depletion of the heat energy gradients.

MEP is consistent with the laws of thermodynamics and has proven to be useful in describing and predicting several geophysical and ecological phenomena. However, no exact analytical understanding of MEP exists, and specification of when the principle is applicable is not precise. The principle applies to sufficiently large or complex systems, or those with sufficient degrees of freedom, but what is sufficient?

It is not the entropy that is produced that is important, Volk and Paulius (2010) argued, but how it is produced. They pointed out that MEP (and irreversible thermodynamics in general) have not shed much light on how entropy is produced and the relative importance of specific processes. While conceding the success in some MEP applications of predicting the overall state of entropy, they showed that two systems with very different internal dynamics could have identical entropy production rates.

In meteorological applications, Liu et al. (2011) argue that MEP provides only a general insight into the overall configuration of the atmosphere. Analysis of negative entropy flow (NEF) provides new insights and superior predictions for weather systems, which they demonstrate via analysis of tropical cyclones. Negentropy flow facilitates the aggregation and convergence of air masses, leading to formation of well-defined low-pressure centers, and sustains the circulation of warm, moist air toward the core of cyclones, fueling intensification. In critiquing the two-box climate model, Lorenz (2010) also argued that the domain of utility of the MEP in climate models is "rather narrow."

Rather than engage the debate over MEP principles (see Martyushev and Seleznev, 2006; Martyushev, 2010; Bruers, 2007; Liu et al. 2011), let us concede that the principle is not a law—but is consistent with laws of thermodynamics. MEP is not applicable or useful to all problems related to thermodynamics in ESS but is demonstrably efficacious in some cases. As Martyushev (2010) pointed out, a principle such as MEP cannot be proved or falsified—rather, observations or experimental results supporting or contradicting it define the regions of applicability. Importantly, where and when MEP does apply, it is a form of thermodynamic selection.

In fluid systems stability is measured in terms of dimensionless parameters such as the Reynolds or Rayleigh number. When these exceed a threshold value, the fluid system is no longer stable and turbulent motions develop. Ozawa et al. (2003) noted that turbulence cannot develop instantaneously, and observation of turbulent motion over a timescale Δt depends on $\Delta t > \tau_{NL}$, where the latter term is a characteristic time constant for turbulent structures to develop in the nonlinear system. Over a sufficiently long time span, then the temperature or velocity differences in the surrounding system decline to negligible levels, the relevant dimensionless parameter declines to 0, and the entire system is in thermodynamic equilibrium. This is the final stage of the nonlinear system and has been referred to as "heat death." The time required for this to occur is τ_{surr}.

This provides some guidance on when MEP is applicable, at least for fluid flows. Using X to represent the relevant parameter (e.g., Prandtl, Rayleigh or Reynolds numbers), and X^* the key threshold value, Ozawa et al. (2003) determined the conditions shown in Table 7.1.

Paltridge's (1978) original two-box model was criticized for lacking a mechanism by which MEP is achieved, and for producing meridional heat fluxes independent of known constraints on atmospheric dynamics. Later Lorenz et al. (2001) showed that equator-to-pole temperature differences on Earth, Titan, and Mars can be produced by a simple two-box model with a MEP selection principle. But how, Jupp and Cox (2010) wondered, could the two-box model reproduce the temperature gradients without reference to atmospheric dynamics or rotation on those planets? They addressed this by extending the two-box model to include atmospheric dynamics on a rotating planet.

A key unrepresented factor in the simplest two-box models is the effect of surface drag on heat and momentum fluxes (and entropy production). Kleidon et al. (2006) addressed this by including surface drag in a series of simulations, concluding that the observed values of drag on Earth corresponded with an MEP state. Jupp and Cox (2010) produced an analysis in between Lorenz et al.'s no-dynamics model and Kleidon et al.'s full numerical dynamics. They developed a set of equations constituting a full description of planetary energy balance, surface-to-atmosphere heat flux, equator-to-pole energy transfer, meridional and zonal dynamics. These are represented, respectively, in the five dimensionless equations below, where t, f, u are non-dimensionalized versions of temperature, energy flux, and wind velocity. The subscripts indicate equator-to-pole difference (ep), flux from surface to atmosphere (a), surface-atmosphere difference (sa), and c_d is a drag coefficient. ξ is a rescaled advection parameter, and τ is wind angle tangent. ω is a measure of planet rotation rate, and ζ is a rescaled rotational parameter. The dimensional equations and derivation of the dimensionless values are given in Jupp and Cox (2010).

Table 7.1 Conditions for maximum energy production in fluid flows

Stability condition	Timescale	Entropy production	MEP Applicable?
$0 < X < X^*$	$\Delta t < \tau_{surr}$	dS/dt = minimum, linear regime	No
$X > X^*$	$\Delta t < \tau_{NL}$; initial stage	Minimum < dS/dt < maximum	No
$X > X^*$	$\tau_{NL} < \Delta t < \tau_{surr}$; developed stage	dS/dt = maximum; nonlinear regime	Yes
$X = 0$	$\Delta t \cong \tau_{surr}$; final stage	0	No

$$t_{ep} + f_a = 1 \tag{7.13}$$

$$\zeta f_a = 2\xi c_d u t_{sa} \tag{7.14}$$

$$2\sqrt{1+\tau^2}\,\zeta f_a = \xi u\left(t_{ep} - 2t_{sa}\right) \tag{7.15}$$

$$c_d u^2 = -\omega\zeta\tau u + \frac{1}{2} \cong \zeta^2\sqrt{1+\tau^2}\left(t_{ep} - 2t_{sa}\right) \tag{7.16}$$

$$\tau c_d u^2 = \omega\zeta u \tag{7.17}$$

When the advective capacity of the atmosphere is sufficiently large and the planetary rotation rate is slow enough—which is the case for Earth and other rocky planets and Saturn's moon Titan in our solar system, Jupp and Cox's model reproduces the MEP state.

In Sect. 4.1.1. it was shown that, if not for the negative feedback of the drag force, higher wind velocities associated with increasing pressure gradients would drive the system to dynamical instability (and vice-versa). Selection favors stability, so the turbulence is maximized in such cases so that negative feedback from drag stabilizes the wind system.

7.4.1 But (How) Is It Selection?

Entropy production and other thermodynamic principles are typically vague with respect to selection. What is meant by "selecting a path"? How is efficiency defined?

In large, complicated systems energy flows can be considered to explore diverse evolutionary paths via random variation or complex interactions. Those that lead to more rapid depletion of gradients survive (and sometimes intensify); those that do not tend to vanish. Thus the varied pathways are selected, for or against. The MEP, for instance, identifies the most probable state, which is generally stable—other development trajectories may occur, but they are less stable.

Thermodynamic selection therefore reflects the VSR criteria: variations in possible trajectories or evolutionary pathways exist; some are selected for or against; and the ones selected for tend to be retained in the short term or to repeatedly recur in the long term.

Bruers and Meysman (2008) demonstrated a correspondence between a resource-consumer chemostat ecosystem model and the Rayleigh-Benard convective fluid system. They found corresponding dynamical equations between the two, as well as comparable physical interpretations. By using criteria of competition and fitness, they were able to correctly calculate the size of convection cells. They also conjectured that "predative behavior" can explain energy dissipation patterns in the fluid system.

From among alternative solutions of the Navier–Stokes and Euler equations, MEP selects the correct physical measure (Glimm et al., 2020). They proved mathematically that MEP is a necessary condition to select the physical solution of the Navier–Stokes and Euler equations for incompressible fluids.

7.5 Other Examples

Multiple examples of explanations of ecosystem evolution, structure, and functioning based on MEP or closely related forms of thermodynamic selection exist (e.g., Patten, 1995; Jorgensen, 1997; Svirezhev, 2000; Ponge, 2012; Stoy et al., 2014; Yen et al., 2014; Lin, 2015; Kleidon, 2023). This section will focus on examples of thermodynamic selection in ESS that are largely or entirely independent of biota.

7.5.1 Estuarine Mixing

Water mixing in estuaries involves the interaction of freshwater inflows, salt water from tides, and (to varying extents) wind and waves. Zhang and Savenije (2018, 2019) examined the thermodynamics of estuarine mixing as a process of energy dissipation. River flow into an estuary from the river has potential energy, which can perform work, relative to the ocean saltwater. From the ocean inland and upstream there is a gradual transition from saline to fresh water as well as an increase in the water level representing an increase in potential energy. Zhang and Savenije (2018, 2019) define alluvial estuaries as systems that can adjust dissipation processes to the energy sources that drive them (i.e., river flow, tides, and to a minor extent in their model, wind and waves). The potential energy of the freshwater is dissipated by mixing.

Zhang and Savenije (2018) assumed that dissipation takes place at maximum power as various mixing mechanisms jointly perform the work. Power is maximized by a dispersion coefficient that represents the combined mixing processes. They developed a differential equation that can be solved in concert with the advection-diffusion equation, which compared well with 52 salinity distributions in 23 estuaries around the world. The relative importance of different mixing processes was not a chief concern; rather the focus was on their combined performance at energy dissipation, which was predicted well by the maximum power approach. This suggests a system responding like a machine approaching the Carnot limit.

Friction is also accounted for in Zhang and Savenije (2019), where they developed an equation for determining the dispersion coefficient D (dimensions of LT^{-1}). The dimensionless coefficient is proportional to the square root of the estuarine Richardson number N_R:

$$\frac{D_g}{\upsilon E} \propto N_r^{0.5} = \sqrt{\left(c_S S \frac{gh}{\upsilon^2} \frac{|Q|T}{AE} \right)} \tag{7.18}$$

υ is tidal velocity, E (m) is the horizontal length scale of gravitational circulation, S is salinity (practical salinity units, psu), c_S is saline expansivity (psu^{-1}), h = depth, Q is freshwater discharge (which can be negative due to tides and backwater effects), T is tidal period (seconds), and A is cross-sectional area. By applying the maximum power assumption Zhang and Savenije obtained

$$D_g(x) = C\sqrt{\frac{S|Q|ET}{B}} \tag{7.19}$$

The subscript g indicates gravitational circulation, x is the longitudinal location along the estuary, and C is a coefficient (psu^{-1} m s^{-1}). Again, they obtained good results in comparison with the observations mentioned above.

Overall, the results indicate thermodynamic selection at work in estuarine mixing.

7.5.2 Fluid Turbulence

Returning to turbulent fluid flow, for the case of a fluid system that exchanges heat and momentum with the surrounding system, Ozawa et al. (2001) proposed that where the fluid system tends to maximize the rate of entropy increase in the surrounding system,

$$\frac{dS_{turb,st}}{dt} = \int \boldsymbol{F}\cdot\nabla\left(\frac{1}{T}\right)dV + \int\frac{\Phi}{T}dV = \int\frac{F}{T}dA = \text{Maximum.} \tag{7.20}$$

$S_{turb,st}$ is the rate of entropy increase due to turbulence (at steady state), **F** is the diabatic heat flux density due to turbulence, T is temperature, V is the volume of the fluid system, and **Φ** represents the rate of kinetic energy dissipation into heat by viscosity, per unit volume of the fluid. F is diabatic heat flux due to turbulence at the boundary (positive outwards), and A is the surface bounding the system from its surroundings.

Ozawa et al. (2001) showed that the proposition of Eq. 7.20 can explain the maximum heat transport for thermal convection, and the maximum momentum transport for turbulent shear flow, which corresponds to the maximum shear working on the system, resulting in maximum viscous dissipation (maximizing **Φ**). They also found that the maximum entropy increase suggested for the global ocean-atmosphere system by the work of Paltridge and others corresponds to Eq. 7.20. A mechanism by which turbulent flow systems adjust to maximum entropy production was proposed by Ozawa et al. (2003). This self-reinforcing mechanism occurs because if a dynamic motion is accelerated, heat transport increases and generates excess available potential energy, further enhancing dynamic motion.

Mihelich et al. (2017) critically re-examined MEP in turbulent flows in several nonequilibrium systems. They compared four optimization principles, based on maximization of energy dissipation, entropy production, and Kolmogorov–Sinai entropy, and minimization of mixing time, and studied the connection between them with simple models describing mixing of a scalar quantity. They found relationships between most probable stationary states of the system using the four principles, providing an empirical justification of the MEP principle in this class of systems, including some turbulent flows, for some boundary conditions, such as those in classical situations that include shear flows, von Karman flows, and Rayleigh-Bénard convection. Otherwise, Mihelich et al. (2017) favor the minimization of the mixing time as a more appropriate principle. However, note that minimum mixing time implies increasing turbulence and entropy production and is thus qualitatively consistent with MEP.

7.5.3 Navier-Stokes Equations

The Navier-Stokes equations describe how the velocity, pressure, temperature, and density of a moving fluid are related. They were derived independently in the early 1800s by M. Navier and G.G. Stokes. The equations are extensions of the Euler equations to include the effects of viscosity on the flow. As shown in Fig. 7.3, the equations are quite complicated, but undergraduate physics and engineering students are taught how to derive them.

While the Navier-Stokes equations could, in theory, be solved analytically, they are too complicated (I mean, look at them!) to make that practical. Solving them using several different approaches is a major task of computational fluid dynamics.

Total entropy of a turbulent fluid flow is the sum of thermal entropy related to fluid density fluctuations, a mixing entropy related to (thermal) concentration gradients, and kinetic energy linked to velocity fluctuations. However, for classical turbulence where fluid density is considered constant, entropy is entirely kinetic and is due to velocity fluctuations. Glimm et al. (2020) presented a theorem directly related

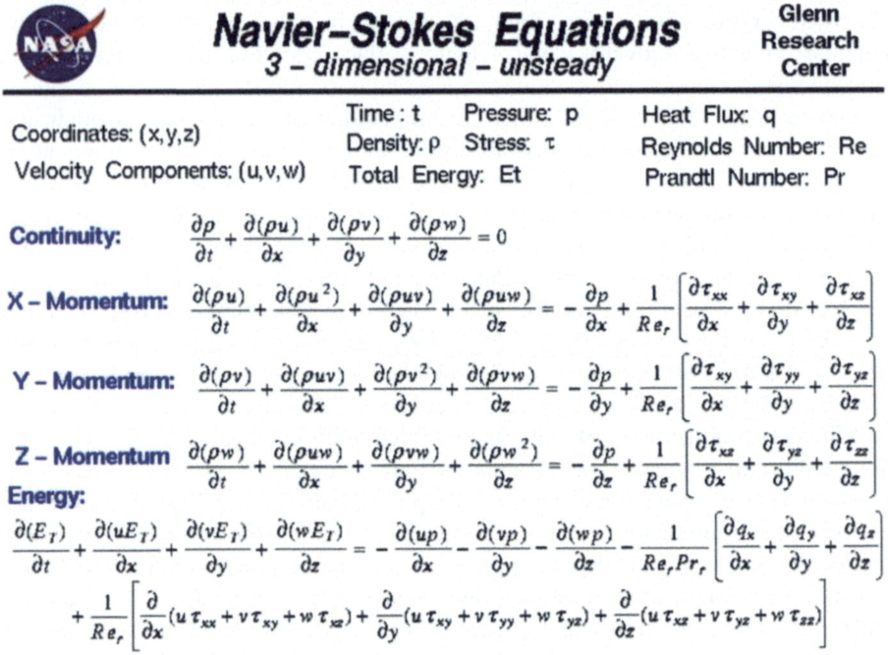

Fig. 7.3 Three-dimensional unsteady-flow version of the Navier-Stokes equations as shown on NASA's "Beginner's Guide to Aeronautics" web page (https://www1.grc.nasa.gov/beginners-guide-to-aeronautics/navier-strokes-equation/; last accessed 29 April 2024)

to selection, that of the available solutions within the velocity configuration state, the one that maximizes entropy production is selected. They proceeded to mathematically prove the theorem. The nonuniqueness of numerical solutions of the Navier-Stokes equations indicates their nonphysical nature, and Glimm et al. (2020) argue that maximum entropy production is a necessary admissibility condition to show that a solution is physically correct.

7.5.4 Evaporation and the Hydrologic Cycle

Evaporation is described by Kleidon (2023) as an example of a process that reflects maximum power—that is, the selection and consequent recurrence of pathways that maximize power, though evaporation is not directly involved in power maximization. At the global scale evapotranspiration balances precipitation, so the strength of the hydrological cycle is strongly but indirectly constrained by thermodynamic optimality. Boundary layer theory is consistent with this interpretation, as maximum power provides a constraint on buoyancy production. Global warming essentially expands (loosens) this constraint, consistent with intensification of the hydrologic cycle.

Maximization of power applies to the sensible heat flux, which is directly linked to maximum buoyancy production. Buoyancy flux is typically written as $\rho\omega'\theta'$, (air density, variation in the vertical wind component, and variation in potential temperature, respectively). Buoyancy production G_p in the total kinetic energy budget is

$$G_b = \frac{g}{\theta}(\rho\omega'\theta'). \tag{7.21}$$

θ is mean potential temperature of the boundary layer, and g is gravitational acceleration.

By integrating over the height of the boundary layer z_b, buoyancy production can be linked to maximum power. This produces the total buoyancy production (units of $W\,m^{-2}$), making it comparable to power obtained from the Carnot limit. Boundary layer height can be represented by the temperature difference between the top of the boundary layer and the surface using the dry adiabatic lapse rate $\Gamma = c_p/g$, so $z_b = c_p/g\ \Delta T$ (c_p is heat capacity of air and ΔT the temperature difference between the surface and the top of the boundary layer). Total buoyancy production is then

$$G_{b,total} = 0.5\,c_p\rho\omega'\theta'\left(\frac{\Delta T}{\theta}\right) = 0.5\,H\left(\frac{\Delta T}{T_s}\right) \tag{7.22}$$

where T_s is an approximation for θ, and $H = c_p\,\rho\omega'\theta'$, the sensible heat flux.

The total buoyancy production expression thus has the form of a thermodynamic Carnot limit of a heat engine powered by sensible heat flux, with the temperature gradient ΔT the driver (Kleidon, 2023).

7.5.5 Climate change and River Dynamics

Eliézer Biao and colleagues (Biao et al., 2015, 2016; Biao, 2017) developed a model based on the LAP to assess climate change effects on the Oueme River, Benin, using a principle of minimum energy expenditure. The key equations are

$$\frac{dZ}{dt} = \psi(q,t) \tag{7.23}$$

$$\frac{d(\lambda Q)}{dt} + \mu Q^{2\mu-1} = \psi(q,t) \Rightarrow \frac{dQ}{dt} = \Upsilon(Q,\psi,\lambda). \tag{7.24}$$

Z describes the variation of the initial state of the catchment; ψ describes the model input, i.e. the rainfall intensity (mm/day), Y describes the model structure, λ and μ are the physical parameters of the model and have no units, Q is river discharge, q is the effective rainfall (mm) and t is the time (day). The recession coefficient λ is a lumped constant representing properties of the watershed, while μ describes the nonlinearity of transformation of rainfall into runoff.

Equation 7.23 describes runoff production, accounting for evapotranspiration and partitioning the remaining precipitation into overland and subsurface flow. Equation 7.24 represents the processes by which the overland and subsurface rainfall volumes are transformed into runoff. Applying the LAP leads to

$$\frac{dQ}{dt} + \frac{\mu}{\lambda} Q^{2\mu-1} = \frac{x(t)}{\lambda} q(t) \tag{7.25}$$

where x describes the wet or dry state of the watershed. The model is structured so that hydrological response to precipitation on day t is reflected at the outlet of the drainage basin on $t + 1$. Discretizing Eq. 7.25,

$$Q_t = Q_{t-1} + \frac{\mu}{\lambda} Q_{t-1}^{2\mu-1} + \frac{x_t}{\lambda} q_{t-1} \tag{7.26}$$

The LAP-based model (HyMoLAP) has been successfully applied and confirmed using empirical data in several subbasins of the Oueme River (Biao et al., 2015, 2016; Biao, 2017).

7.5.6 ENSO and the Global Energy Budget

El Niño, a climate phenomenon characterized by the warming of sea surface temperatures in the central and eastern tropical Pacific Ocean, plays a significant role in the global energy balance. During El Niño, warm waters accumulate in the eastern Pacific and radiate heat into the atmosphere. *The Guardian* newspaper called El Niño a "rechargeable heat battery," accumulating heat in the ocean and releasing it to other regions during an El Niño event. Solar radiation inputs were the major driver of these events until recent decades. Recent studies indicate that anthropic global warming now plays a major role.

During El Niño, trade winds in the equatorial Pacific weaken, pushing warm water east toward the west coast of South America. La Niña events have the opposite effects. El Niño and La Niña represent warm and cold phases of the El Niño-Southern Oscillation (ENSO) cycle, with the Southern Oscillation representing the atmospheric components. The latter involves east-west fluctuations in atmospheric pressure over the tropical Pacific. ENSO events have major global weather, climate, and oceanographic effects. Here we focus on the role of ENSO in the global energy budget.

The tropics receive the most solar radiation on Earth, and general oceanic and atmospheric circulations redistribute that energy from net sources to net sinks. Interannual variability plays a role in this redistribution, and ENSO is the dominant source of interannual variability in the tropical Pacific. The western Pacific warm pool (Fig. 7.4) accumulates oceanic heat, and El Niño events discharge this heat toward the eastern Pacific. This leads to relative warming in the central and eastern and cooling in the western Pacific. Warming in the central-eastern ocean triggers evaporative cooling, transforming the ocean heat to latent heat which is released when condensation occurs. This warms the atmospheric column over the central/eastern Pacific and weakens the Walker circulation (Fig. 7.5). The latter describes tropospheric air flow in the tropics and is driven by differences in heat distribution between land and sea. Normally Walker circulation is characterized by easterly surface winds in the western and westerly surface winds in the eastern Pacific. El Niño weakens the easterly trade winds, while La Niña intensifies them.

Weakening of the Walker circulation results in an anomalous divergence of horizontal transport of moist static energy (MSE) in the atmosphere above the central and eastern Pacific. The energy from this oceanic heat pump is redistributed around the globe chiefly by El Niño-related modifications to atmospheric circulation.

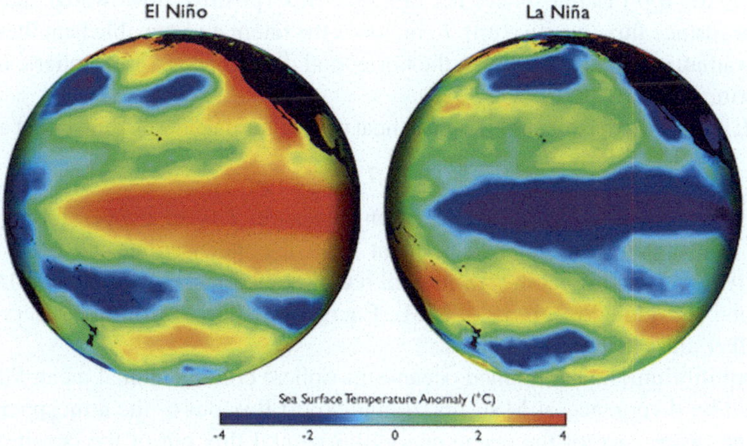

Fig. 7.4 Typical sea surface temperature anomalies during El Niño and La Niña events. (Reprinted from https://www.climate.gov. © 2024, NOAA)

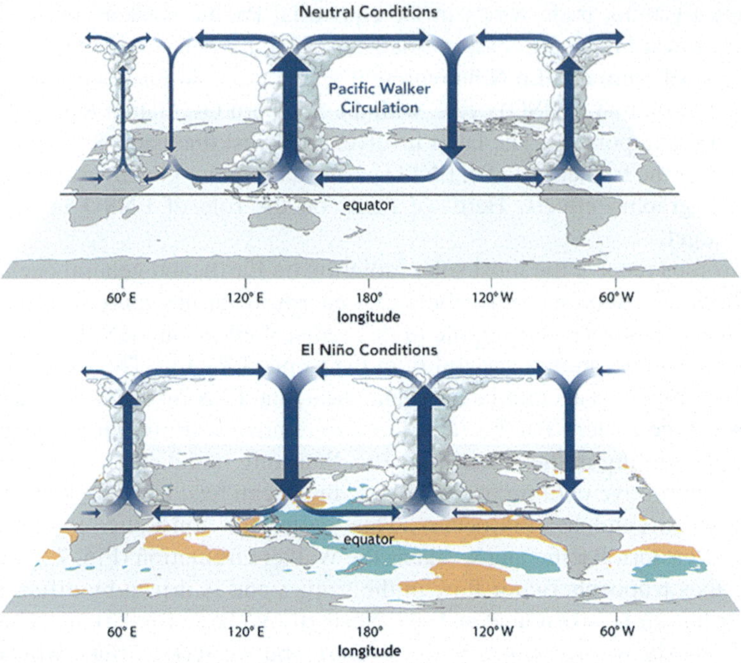

Fig. 7.5 The Walker circulation during El Niño and normal or neutral conditions. (Reprinted from https://www.climate.gov. © 2024, NOAA)

In a column of the atmosphere the deficit or excess of energy is given by Q, the diabatic heating:

$$Q = Rad_{TOA} + F_s \tag{7.27}$$

Rad_{TOA} is the top-of-atmosphere net radiative flux (positive downward) and F_s the total net surface flux (positive up). F_s includes the latent and sensible heat fluxes and the net radiative flux. Q is equal to the horizontal divergence of atmospheric MSE at equilibrium.

MSE combines the dry static energy (heat plus potential energy) and latent energy:

$$MSE = C_P T + gz + L_q \tag{7.28}$$

C_p is the specific heat of air at constant pressure, T is air temperature, g is the gravity constant, z is altitude, L is latent heat of condensation, and q is specific humidity. The western Pacific warm pool receives the most energy from TOA insolation and is a global reservoir of MSE. Energy is exported from the reservoir via the Walker and Hadley circulations.

At equilibrium, in a combined ocean-atmosphere column annual mean Rad_{TOA} is balanced by divergence of MSE horizontal export flux out of the atmospheric column, and divergence of the ocean energy horizontal flux out of the ocean column ($\underset{\nabla}{\rightarrow} F_{MSE}$, $\underset{\nabla}{\rightarrow} F_O$, respectively).

$$RAD_{TOA} = \underset{\nabla}{\rightarrow} F_{MSE} + \underset{\nabla}{\rightarrow} F_O \tag{7.29}$$

As summarized by Saint-Lu et al. (2016): eastern Pacific warming delivers heat to the atmosphere via latent heat fluxes and radiative longwave fluxes. These are amplified by longwave cloud feedback. The atmospheric energy is then exported from the Pacific, mainly towards the Indian and Atlantic oceans areas. The RAD_{TOA} anomalies during El Niño, the discharge from the warm pool, and the MSE export show that El Niño events play a key role in global energy redistribution by exporting some of the energy excess from the tropical Pacific.

Selection principles derived from the Second Law of Thermodynamics and the maximum entropy production principle are based on depletion of gradients as rapidly as possible and accomplishing this as efficiently as possible, according to the LAP. The surplus of heat energy in the western Pacific warm pool creates strong gradients of temperature (and sea surface height). The export of this heat toward the east, by itself, does not constitute a maximum-power gradient depletion mechanism. However, the resulting redistribution beyond the equatorial Pacific by El Niño does potentially represent such a mechanism. If this is the case, El Niño events would be expected to be weaker and to have reduced influence on global energy redistribution when radiative forcing reduces the energy in the tropics. Saint-Lu et al. (2016) found this to be the case, using simulations comparing the pre-Industrial condition (to eliminate effects of anthropic warming) to the middle Holocene about 6 Ka, when insolation brought less annual energy to the tropics and more to higher latitudes, reducing the gradients to be depleted.

Cheng et al. (2019) looked in greater detail at the changes in ocean heat storage associated with ENSO and identified several mechanisms associated with the heat redistribution. This includes the aforementioned changes in the Walker circulation, variations in the shape of the ocean thermocline, enhanced sea to air heat fluxes driven by higher sea surface temperatures, and local adiabatic redistribution of heat both vertically and laterally in the tropical Pacific and Indian oceans.

The Indonesian throughflow, a teleconnection driven by ENSO, warms the Atlantic and Indian Oceans during El Niño. The sea surface of the far western Pacific is generally warmer and higher in elevation than the adjacent area of the Indian ocean, driving upper thermocline water down-gradient through several passages in the Indonesian archipelago. This, the only pathway for marine exchange between the Pacific and Indian Oceans in the tropics, is responsive not only to ENSO, but also to the Indian Ocean Dipole (an irregular oscillation of sea surface temperatures between eastern and western Indian Oceans; Sprintall & Révelard, 2014).

ENSO is characterized by complex nonlinear dynamics, dense interconnections with other atmospheric modes and phenomena, and interactions among several different layers of the atmosphere and ocean. These complexities are still being untangled, such as Hogikyan et al.'s (2022) investigation of intense tropical tropospheric warming in response to El Niño. But consideration of ENSO's role in global energy redistribution and its relationship to maximum power and MEP is a promising avenue for improving understanding (e.g. Saint-Lu et al., 2016; Cheng et al., 2019). Some systems maximize power overall by variations, so they are not necessarily operating at maximum power at every location all the time. For instance, midlatitude synoptic dynamics overall apparently maximize power with respect to poleward heat transport (Kleidon, 2023), but it would probably be fruitless (at least in this context) to analyze power and dissipation for each low- or high-pressure system. An intriguing possibility is that ENSO is broadly analogous to diurnal cycles, where power is performed in daylight by dry convection. This charges the lower atmosphere with heat, while overnight, no power is performed while cooling and discharging the added heat by emission to space. In this analogy, ENSO cycles between warm (El Niño), cold (La Niña), and neutral phases recharge and discharge the tropical heat battery.

7.6 Thermodynamic Selection: Summary

The laws of thermodynamics, especially the second law, underpin the least action, maximum power, and maximum entropy production principles. They boil down to depleting energy gradients as rapidly and efficiently as possible. As varied forms and pathways of energy flows and transformations "explore" their possibility space, those that achieve the gradient depletion and efficiency criteria are retained in the short term and recur repeatedly over time. The explorations that do not meet these criteria tend to fade and vanish and do not show a pattern of recurrence. These principles can be grouped together as forms of thermodynamic selection.

Thermodynamic selection, often in the form of optimality principles, have frequently been applied in ecology and ecosystem science. There is also a tradition of application in geophysics and abiotic phenomena. Thermodynamic selection occurs in general processes such as turbulent fluid flows and atmospheric energy exchanges. These are manifested in global phenomena such as atmospheric and ocean general circulation, evaporation and the hydrologic cycle, and teleconnection patterns such as ENSO. Thermodynamic selection is also evident at more local scales, e.g. in fluvial and estuarine processes.

References

Annila, A. (2021). Statistical physics of evolving systems. *Entropy, 23*, 1590. https://doi.org/10.3390/e23121590

Annila, A., & Salthe, S. (2010). Physical foundations of evolutionary theory. *Journal of Non-Equilibrium Thermodynamics, 35*, 301–321. https://doi.org/10.1515/JNETDY.2010.19

Biao, E. I. (2017). Assessing the dynamics of climate change on river discharge dynamics in Oueme River basin (Benin, West Africa). *Hydrology, 4*, 47. https://doi.org/10.3390/hydrology4040047

Biao, E. I., Gaba, C. O. U., Alamou, A. E., & Afouda, A. A. (2015). Influence of the uncertainties related to the random component of rainfall inflow in the Oueme river basin (Benin, West Africa). *International Journal of Current Engineering Technology, 5*, 618–1629.

Biao, E. I., Alamou, A. E., & Afouda, A. A. (2016). Improving rainfall-runoff modelling through the control of uncertainties under increasing climate variability in the Oueme River basin (Benin, West Africa). *Hydrological Sciences Journal, 61*, 2902–2915.

Bruers, S., Meysman, F. (2008). A useful correspondence between fluid convection and ecosystem operation. arXiv physics-fluid dynamics, arXiv:0708.0091.

Budyko, M. I. (1969). The effect of solar radiation variations on the climate of the earth. *Tellus, 21*, 611–619.
Cheng, L., Trenberth, K. E., Fasullo, J. T., et al. (2019). Evolution of ocean heat content related to ENSO. *Journal of Climate, 32*, 3529–3556.
Cushman, S. A. (2023). Entropy, ecology and evolution: Toward a unified philosophy of biology. *Entropy, 25*(405). https://doi.org/10.3390/e25030405
Dewar, R. C. (2009). Maximum entropy production as an inference algorithm that translates physical assumptions into macroscopic predictions: Don't shoot the messenger. *Entropy, 11*, 931–944. https://doi.org/10.3390/e11040931
Dyke, J. G., & Kleidon, A. (2010). The maximum entropy production principle: Its theoretical foundations and applications to the earth system. *Entropy, 12*, 613–630. https://doi.org/10.3390/e12030613
Glimm, J., Lazerev, D., Ghen, G.G. 2020. Maximum entropy production as a necessary admissibility condition for the fluid Navier–Stokes and Euler equations. SN Applied Sciences 2, 2160. https://doi.org/10.1007/s42452-020-03941-2.
Hogikyan, A., Resplandy, L., & Fueglistaler, S. (2022). Cause of the intense tropics-wide tropospheric warming in response to El Niño. *Journal of Climate, 35*, 2933–2944. https://doi.org/10.1175/JCLI-D-21-0728.1
Huang, H. Q., & Nanson, G. C. (2000). Hydraulic geometry and maximum flow efficiency as products of the principle of least action. *Earth Surface Processes and Landforms, 25*, 1–16.
Huang, H. Q., Chang, H. H., & Nanson, G. C. (2004). Minimum energy as a general form of critical flow and maximum flow efficiency and for explaining variations in river channel pattern. *Water Resources Research, 40*, W04502.
Huang, H. Q., Deng, C., Nanson, G. C., et al. (2014). A test of equilibrium theory and a demonstration of its practical application for predicting the morphodynamics of the Yangtze River. *Earth Surface Processes and Landforms, 39*, 669–675.
Hutchinson, G. E. (1957). Concluding remarks. *Cold Spring Harbor Symposium on Quantitative Biology, 22*, 415–427.
Jorgensen, S. E. (1997). *Integration of ecosystem theories: A pattern*. Kluwer. 388p. 3rd ed., 2002.
Jupp, T. E., & Cox, P. M. (2010). MEP and planetary climates: Insights from a two-box climate model containing atmospheric dynamics. *Philosophical Transactions of the Royal Society B, 365*, 1355–1365. https://doi.org/10.1098/rstb.2009.0297
Kleidon, A. 2021. Physical limits of wind energy within the atmosphere and its use as renewable energy: From the theoretical basis to practical implications, Meteorologische Zeitschrift 30, 203–225, https://doi.org/10.1127/metz/2021/1062, 2021a.
Kleidon, A. (2023). Working at the limit: A review of thermodynamics and optimality of the earth system. *Earth System Dynamics, 14*, 861–896.
Kleidon, A., Fraedrich, K., Kirk, E., & Lunkheit, F. (2006). Maximum entropy production and the strength of boundary layer exchange in an atmospheric general circulation model. *Geophysical Research Letters, 33*, L06706. https://doi.org/10.1029/2005GL025373
Lin, H. (2015). Thermodynamic entropy fluxes reflect ecosystem characteristics and succession. *Ecological Modelling, 298*, 75–86.
Liu, Y., Liu, C., & Wang, D. (2011). Understanding atmospheric behaviour in terms of entropy: A review of applications of the second law of thermodynamics to meteorology. *Entropy, 13*, 211–240. https://doi.org/10.3390/e13010211
Lorenz, R. D. (2010). The two-box model of climate: Limitations and applications to planetary habitability and maximum entropy production studies. *Philosophical Transactions of the Royal Society B, 365*, 1349–1354. https://doi.org/10.1098/rstb.2009.0312
Lorenz, R. D., Lunine, J. I., Withers, P. G., & McKay, C. P. (2001). Titan, mars and earth: Entropy production by latitudinal heat transport. *Geophysical Research Letters, 28*, 415–418. https://doi.org/10.1029/2000GL01233

Lotka, A. J. (1922). Contribution to the energetics of evolution. *Proceedings of the National Academy of Sciences (USA), 8*, 147–151.
Martyushev, L. M. (2010). The maximum entropy production principle: Two basic questions. *Philosophical Transactions of the Royal Society B, 365*, 1333–1334. https://doi.org/10.1098/rstb.2009.0295
Martyushev, L. M., & Seleznev, V. D. (2006). Maximum entropy production principle in physics, chemistry and biology. *Physics Reports, 426*, 1–45.
Michaelian, K. (2011). Thermodynamic dissipation theory for the origin of life. *Earth System Dynamics, 2*, 37–51. https://doi.org/10.5194/esd-2-37-2011
Michaelian, K. (2023). The non-equilibrium thermodynamics of natural selection: From molecules to the biosphere. *Entropy, 25*, 1059.
Mihelich, M., Faranda, D., Paillard, D., & Dubrulle, B. (2017). Is turbulence a state of maximum energy dissipation? *Entropy, 19*, 154. https://doi.org/10.3390/e19040154
Nanson, G. C., & Huang, H. Q. (2008). Least action principle, equilibrium states, iterative adjustment and the stability of alluvial channels. *Earth Surface Processes and Landforms, 33*, 923–942.
Nanson, G. C., & Huang, H. Q. (2017). Self-adjustment in rivers: Evidence for least action as the primary control of alluvial channel form and process. *Earth Surface Processes and Landforms, 42*, 575–594.
Odum, H. T., & Pinkerton, R. C. (1955). Time's speed regulator: The optimum efficiency for maximum power output in physical and biological systems. *American Scientist, 43*, 331–342.
Ozawa, H., Shimokawa, S., & Sakuma, H. (2001). Thermodynamics of fluid turbulence: A unified approach to the maximum transport properties. *Physical Review E, 64*, 026303.
Ozawa, H., Ohmura, A., Lorenz, R. D., & Pujol, T. (2003). The second law of thermodynamics and the global climate system: A review of the maximum entropy production principle. *Reviews of Geophysics, 41*. https://doi.org/10.1029/2002RG000113
Paltridge, G. W. (1978). The steady-state format of global climate. *Quarterly Journal of the Royal Meteorological Society, 104*, 927–945.
Patten, B. C. (1995). Network integration of ecological extremal principles: Exergy, emergy, power, ascendancy, and indirect effects. *Ecological Modelling, 79*, 75–84.
Phillips, J. D. (2008). Goal functions in ecosystem and biosphere evolution. *Progress in Physical Geography, 32*, 51–64.
Ponge, J.-F. (2012). Disturbances, organisms and ecosystems: A global change perspective. *Ecology and Evolution, 3*, 1113–1124.
Saint-Lu, M., Braconnot, P., Leloup, J., & Marti, O. (2016). The role of El Niño in the global energy redistribution: A case study in the mid-Holocene. *Climate Dynamics, 52*, 7135–7152. https://doi.org/10.1007/s00382-016-3266-5
Skene, K. R. (2015). Life's a gas: A thermodynamic theory of evolution. *Entropy, 17*, 5522–5548. https://doi.org/10.3390/e17085522
Sprintall, J., & Révelard, A. (2014). The Indonesian throughflow response to Indo-Pacific climate variability. *Journal of Geophysical Research: Oceans, 119*, 1161–1175. https://doi.org/10.1002/2013JC009533
Stoy, P. C., Lin, H., Novick, K. A., et al. (2014). The role of vegetation on the ecosystem radiative entropy budget and trends along ecological succession. *Entropy, 16*, 3710–3731. https://doi.org/10.3390/e16073710
Svirezhev, Y. M. (2000). Thermodynamics and ecology. *Ecological Modelling, 132*, 11–22.
Ueltzhöffer, K., Da Costa, L., Cialfi, D., & Friston, K. (2021). A drive towards thermodynamic efficiency for dissipative structures in chemical reaction networks. *Entropy, 23*, 1115. https://doi.org/10.3390/e23091115
Volk, T., & Paulius, O. (2010). It is not the entropy you produce, rather, how you produce it. *Philosophical Transactions of the Royal Society B, 365*, 1317–1322. https://doi.org/10.1098/rstb.2010.0019

Yen, J. D. L., Paganin, D. M., Thomson, J. R., & MacNally, R. (2014). Thermodynamic extremization principles and their relevance to ecology. *Austral Ecology, 39*, 619–632.
Zhang, Z., & Savenije, H. H. G. (2018). Thermodynamics of saline and fresh water mixing in estuaries. *Earth System Dynamics, 9*, 241–247. https://doi.org/10.5194/esd-9-241-2018
Zhang, Z., & Savenije, H. H. G. (2019). Maximum power of saline and fresh water mixing in estuaries. *Earth System Dynamics, 10*, 667–684. https://doi.org/10.5194/esd-10-667-2019

Geophysical Selection and Earth Surface System Evolution

8

Abstract

The different forms of geophysical/abiotic, biological, and ecological selection interact and overlap. This chapter presents examples of the effects of selection in landscape evolution. These include the coevolution of hydrological, soil, geomorphological, and vegetation systems, tied together in this case via percolation theory, and the water balance, where selection is demonstrated via multiple approaches, including Budyko curves, percolation theory, vegetation optimality, the proportionality hypothesis, and maximum entropy production. Landscape evolution via emergent selection is also shown to be a form of adaptation, using an analysis of preferential flow and "store and pour" structures in hydrological systems. Region-specific case studies include the White Desert, Egypt; fluvial-estuary transition zones in the U.S. Atlantic Coastal Plain; and the Brazilian Cerrado. Efficiency selection consistently operates in landscapes, but a progression toward greater efficiency is not always observed. This is due to the probabilistic and local nature of selection, and changes in dominant controls of landscape evolution, and therefore of selection criteria, over time. Energy dissipation in one process may decrease efficiency in another. Selection operates on multiple components simultaneously, with effects that may counteract each other.

8.1 Abiotic Selection, Biotic-Abiotic Interactions, and Evolution

We've already seen how different types of selection in physical processes in Earth surface systems (ESS) overlap and interact. The main point of identifying different types or forms of selection is to reflect the variety of selection phenomena and organize review and discussion, not to definitively classify all geophysical selection as one type or another. Weathering, erosion, and sediment transport, for example,

J. D. Phillips, *Abiotic Selection in Earth Surface Systems*, Geophysics and Environmental Physics, https://doi.org/10.1007/978-3-031-85862-8_8

involve both gradient and resistance selection. Network and thermodynamic selection are intertwined. And so on.

As ESS and landscapes evolve, selection operates in multiple processes acting simultaneously, in some cases semi-independently, sometimes offsetting or sometimes reinforcing other selection processes. As we have seen, geophysical selection operates at time scales ranging from those of fluid dynamics to those of geological evolution. Given the complex interactions involved in landscape evolution and the vast range of scales at which selection occurs, it is difficult to provide a comprehensive global overview or synthesis on the subject. However, some illustrative examples can serve to identify common threads and general principles.

8.2 Backbones of Landscape Evolution

We encountered percolation theory in Chap. 6, where the focus was on selection of optimal network pathways in water movement and related fluxes through soil and rock. As mentioned there, percolation has been used to tie together several hydrological, geomorphological, soil, and ecological phenomena, and that is the focus here. Allen G. Hunt and colleagues have led the way in this work.

While percolation theory applications to ESS began much earlier, a major advance was reported by Hunt (2016), linking soil/groundwater hydrology, soil formation, and vegetation growth, using critical path analysis (CPA). The latter is perhaps more familiar in engineering, where it is a method to identify the longest sequence of tasks in a project, which is a constraint on the minimum time to complete the project. In porous media, CPA isolates the role of pore-size distributions in characterizing fluid flows. In heterogeneous media such as rock, soil, and sediment, CPA is used in conjunction with percolation theory (PT) to analyze the effects of connectivity along with the distribution of pore sizes, linking local properties of pore size to larger scale behaviors of fluid flow.

Where CPA can be used to calculate hydraulic conductivity (*K*), Hunt argued, advective transport scaling relationships from percolation theory describe solute transport. Hunt (2016) presented two solute transport relationships, used to predict soil development and edaphic constraints on natural plant growth. Hunt used the known exponents from percolation theory (see Sect. 6.3) and characteristic subsurface flow velocities. Flow velocity constrains the optimal plant growth rates, and the percolation scaling relationship constraining vegetation growth was shown to be consistent with empirical data across time scales of hours to 100 ka. Based on comparison with soil development data from 23 separate studies, the scaling function matches time scales for formation of soils from years to hundreds of millions of years. The unifying factor is the common origin of the time and space coordinates in the transport time through a single pore (of roughly micrometer size) at a typical subsurface pore-scale flow velocities. Hunt (2016) attributed the difference in evolutionary time scales to the hierarchical nature of vascular plant root systems, which accelerates access to nutrients compared to rates of physical transport in the soil.

As indicated in Eq. 6.18, transport time t through the system is proportional to system size x: $t = t_o(x/x_o)^{Db}$, where t_o, x_o are characteristic time and length scales, and D_b the fractal dimension of the percolation backbone. Length scale is taken as pore size, assumed to be commensurate with heterogeneity associated with particle size, and a single pore is the maximum scale where the advection-dispersion equation is applicable (Hunt et al., 2011). In Hunt (2016) this $x_o = 1$ μm, though he acknowledges that larger sizes (e.g., 10 μm for vegetation) might be more appropriate in some cases. Typical soil and groundwater flow velocities are on the order of 10^{-6} m s^{-1}, so t_o was set to 1 s. Then

$$x = x_o \left(\frac{t}{t_o} \right)^{1/D_b} = 1\mu m \left(\frac{t}{1s} \right)^{0.53} \tag{8.1}$$

The first part of Eq. 8.1 is the same as Eq. 6.18. This gives estimated time scales of 10^4–10^5 years for a meter of transport; too slow for biota to rely on random percolation. The Peclet number for mass transfer is $Pe = uL/D$, where u is flow velocity, L is length, and D is mass diffusivity. Based on published diffusion constants for solutes in porous media, Hunt estimated that $Pe > 1$ at about 10 μm, indicating advection dominating over diffusion.

Assuming a connection between water and solute movement and soil development, a soil production function is given (neglecting any erosion or removal) by

$$\frac{dx}{dt} = 0.53 \left(\frac{1\mu m}{1s} \right) \left(\frac{x}{x_o} \right)^{-0.87} \tag{8.2}$$

where x in this case is soil depth. The exponents in Eqs. 8.1 and 8.2 are derived from the percolation backbones (Sect. 6.3), indicating efficiency selection for optimum pathways.

Turning to plants, roots, and nutrient acquisition, the hierarchical (branching dendritic) structure of root systems is the single most important factor in increasing nutrient availability. The need for roots to find the paths with lowest energy expenditure (which tends to increase nutrient fluxes) should, Hunt argues, guide their development to optimal paths from PT. Arguing that root systems are often concentrated in a relatively thin planar layer of soil, Hunt (2016) applied the 2D optimal path exponent $D_o = 1.21$, leading to a version of Eq. 8.1 with an exponent of 0.82. The simple formula reflects effects of both hydraulic and nutrient limitations on growth because the time scale, t_0, is determined by soil hydraulic conductivity. The three-dimensional optimal path exponent of 1.43 would lead to the replacement of 0.82 by 0.7.

The results reflect the coevolution of hydrology, soils, and vegetation governed (or at least influenced) by selection for optimal flow or growth paths. There is considerable scatter in the global datasets used to test the PT/CPA analysis, but enough agreement, and constraints on the results, to support Hunt's (2016) conclusions. In Hunt (2017) the analysis was extended to produce explicit predictions, and Hunt et al. (2021) linked subsurface flow, weathering, soil formation, plant growth, and long-term drainage basin evolution.

Figure 8.1, originally from Renschler (2000) via Loague and Corwin (2006), was a major catalyst in Hunt's thinking. Hunt (2016; Hunt et al. 2021) added the strips to indicate important spatiotemporal scaling relationships of Eq. (8.1). The width of the strips indicates approximately 2 orders of magnitude variability in mean annual flow rates, which is largely due to climate-driven variations in water supply, but also the heterogeneity of soil properties (Hunt et al., 2021). They also added a symbol for mineral dissolution and moved the position of the landscape evolution icon to represent a weathered depth. The lower strip shows radial extent of roots (and plant height) as a function of time, with a slope of 0.83. The upper strip, with a slope of 0.53, indicates soil depth as a function of time. These <1 slopes suggest that soil depth and the physical magnitude of plant depths have decreasing growth rates over time. The exponents governing these scaling relationships are the optimal path exponent in two dimensions for vegetation growth and the percolation backbone in three dimensions for soil formation.

Based on the assumption that soil formation with respect to depth is constrained by vertical solute transport rates (supported by Egli et al., 2018), the vertical positioning of the lower strip was selected to make the ranges of solute and flow velocities equivalent at a length scale corresponding to a typical silt-sized grain of about

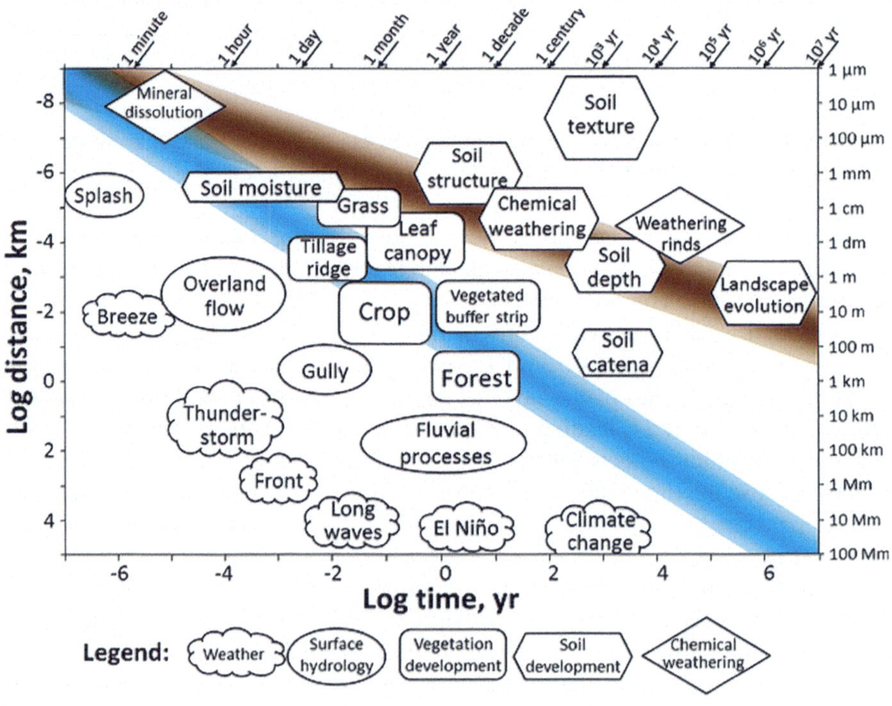

Fig. 8.1 Processes and their characteristic length and time scales. The shaded regions, which link processes across scales, correspond to ranges of fluid and solute velocities expected in the subsurface. (Reprinted under CC-BY-4.0 license from Hunt et al., 2021. © 2021, The Author(s))

30 μm (and timescale of around 30 s). The reduction of solute velocity with increasing scale therefore commences at the fundamental network scale of a single grain.

A series of other papers further explored the links among subsurface flow, solute transport, weathering, soil formation, and root growth, and extended the arguments to ecosystem net primary productivity, all linked by the common selection process of optimal flow or growth paths along the routes of minimal resistance predicted by PT (Yu & Hunt, 2017; Yu et al., 2017; Hunt et al., 2020, 2023a, b, 2024). In Hunt (2024) he tied this body of work to the water balance, and the development of hydrology more broadly.

8.3 The Water Balance

Over the long term, it is sometimes assumed that water storage as soil and groundwater is in steady state (inputs ≈ outputs), so that storage changes are negligible (which is clearly not the case at seasonal and shorter time scale). Therefore, for the long-term average annual water balance at the watershed scales, precipitation (P) inputs can be partitioned into evapotranspiration (ET) and runoff (Q), all with dimensions of LT^{-1} such as mm year^{-1}.

$$P = ET + Q \tag{8.3}$$

Under these assumptions Mikhail Budyko (1974) developed a conceptual framework linking the water and energy balances, and more generally connecting the atmosphere, hydrosphere, and biosphere. Radiation (R_n; mean annual net radiative heat flux from the atmosphere to the land surface) is the product of the latent heat of evaporation (L), evapotranspiration, and sensible heat flux (H) from the land surface into the atmosphere:

$$R_n = L \times ET \times H, \tag{8.4}$$

with R_n, $(L{*}ET)$, and H all having dimensions of MT^{-3}.

Equations 8.3 and 8.4 imply that (at these spatiotemporal scales) all precipitation eventually leaves the watershed as runoff or evapotranspiration, and insolation is not stored as heat but leaves the land surface as latent heat or via heat flux back to the atmosphere. As ET = latent heat flux, Budyko postulated an energy-balance constraint on the water balance:

$$\frac{R_n}{L} = ET + \frac{H}{L} \tag{8.5}$$

Therefore, maximum possible evapotranspiration (ET_o) occurs when "green water" available for ET is not limiting, sensible heat flux is negligible, and all incoming solar radiation is used to convert liquid water to water vapor. Budyko conjectured that $ET = f(P, ET_o)$ over sufficiently long time scales. Note that by incorporating changes in soil moisture storage, this concept is compatible with seasonal-scale water balance frameworks, with ET, ET_o, respectively, corresponding to actual and potential evapotranspiration.

The ET-P relationship, in Budyko's framework, is limited by wet and dry conditions, respectively.

$$ET \rightarrow ET_o \, as \, P \rightarrow \infty \tag{8.6}$$

$$ET \rightarrow P \, as \, ET_o \rightarrow \infty \tag{8.7}$$

Wet conditions are energy-limited by ET_o, while dry conditions are water-limited by *P.* Eqs. 8.6 and 8.7 constitute the Budyko hypothesis, leading to the Budyko equation and the Budyko curve, which are iconic in hydrology. Sposito (2017) reviews Budyko's framework and subsequent developments to it. The Budyko equation is typically expressed as

$$\frac{E}{P} = \varphi\left(\frac{E_o}{P}\right) \tag{8.8}$$

where the function φ represents the Budyko curve. The curve is a graph with the ratio of actual to *ET* to *P* on the y-axis and E_o/P on the x-axis, called the aridity index. Figure 8.2 shows an example, with the energy- and water-limited lines shown along with fitted data. The inflection in the limit line shows the distinction between energy and water limitations at an aridity index of 1.

Sposito (2017) showed that all versions of the Budyko equation can be derived from an assumption that $f(ET_o, P)$ is a homogeneous function of its arguments, such that if the independent variables ET_o and *P* are multiplied by the same arbitrary factor λ, the value of $f(ET_o, P)$ is also multiplied by λ:

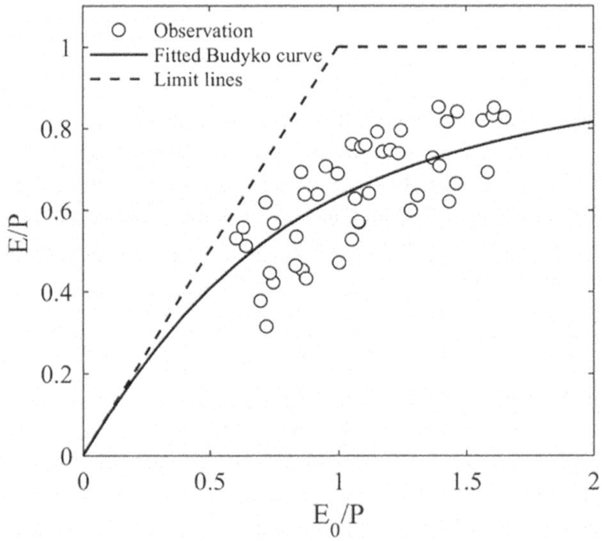

Fig. 8.2 An example of a Budyko curve for data from sub-catchments of the Huaihe River basin, China. (Adapted from Qui et al. (2018). © The Author(s) under CC-BY-4.0 license)

$$f(\lambda ET_o, \lambda P) = \lambda ET \tag{8.9}$$

This is mathematically equivalent to

$$ET = P\,f\left(\frac{ET_o}{P}\right) \tag{8.10}$$

8.3.1 Budyko Curve and Vegetation Optimality

Where storage is indeed a negligible contributor to the long term mean water balance, the Budyko limits to ET are defined by conservation of energy (ET ≤ mean annual net radiation) and mass (ET ≤ mean annual precipitation). In many cases measured values plot relatively close to these limits. Inspired by this, empirical methods were developed to estimate mean annual Q and ET from measured net radiation and precipitation. Because parametric versions of the Budyko equation do not necessarily represent biophysical processes (Sposito, 2017; Reaver et al., 2022), the convergence of catchments around such an empirical curve and the apparent self-organization has been difficult to explain. The vegetation optimality model (VOM) of Nijzink and Schymanksi (2022) suggests that by plants use mechanisms such as, e.g., rooting depths, stomatal control, and vegetation cover to maximize net carbon profit (the difference between uptake from the atmosphere and C used to maintain vegetation tissue). They show that the VOM produces results relatively close to a Budyko curve, suggesting that the convergence of many catchments toward the curve can be explained by selection for maximization of net carbon profit.

Hunt et al. (2021) produced an analytical formula very similar to that of Nijzink and Schymanski (2022) based on optimization of net primary productivity (NPP) relative to the hydrologic fluxes Q and ET. The latter is the single best predictor of NPP (Rosenzweig, 1968):

$$NPP = c(ET)^s \tag{8.11}$$

where c is an empirical constant. Rosenzweig (1968) estimated $s = 1.69$, which Hunt et al. (2017) revised to $s = 1.9$ based on the 2D fractal dimensionality of pore networks from percolation theory. The latter is a good predictor of data up to a maximum NPP determined by light limitation (Hunt et al., 2017, 2024).

Reasoning that plants could not grow where ET = 0, and there would be no medium for plant growth if plants used all moisture inputs (ET = P), as water input is necessary for soil formation, there must be a maximum productivity when ET is a fraction of P greater than zero and less than 1: $ET = aP$, $0 < a < 1$. Using universal exponents from percolation theory, $D_b = 1.87$ (the fractal dimension of the percolation cluster), and $D_f = 1.9$ (for root systems, based on the dimension of a sample-spanning percolation cluster),

$$a = \frac{D_f}{D_f + 1/(D_b - 1)} = 0.623 \tag{8.12}$$

Depending on the aridity index,

$$ET = aE_o, \quad \frac{E_o}{P} < 1 \tag{8.13}$$

$$ET = P\left[1 - (1 - a)P / E_o\right], \quad \frac{E_o}{P} > 1 \tag{8.14}$$

Equations 8.13 and 8.14 were used in another study by Hunt et al. (2023) for predicting changes in streamflow. Substituting these into the power function for NPP (with $a = 0.623$),

$$NPP = c(0.623E_o)^{1.9} \quad \frac{E_o}{P} < 1 \tag{8.15}$$

$$NPP = c\left[P - 0.377\left(P^2 / E_o\right)\right]^{1.9} \quad \frac{E_o}{P} > 1 \tag{8.16}$$

Note that the optimization in Hunt et al. (2024) is applied at the ecosystem level rather than individual plants.

8.3.2 Proportionality Hypothesis and Maximum Entropy Production

The proportionality hypothesis (PH) is that the ratio of ET to its potential value is proportional to the value of Q to its potential value. Wang et al. (2015) argued that the Budyko curve can be conceptualized as the result of applying the PH to estimate mean annual water balance. The proportionality hypothesis, in turn, can be derived from maximum entropy production and thereby linked to selection.

The hypothesis can be traced to the SCS curve number method for runoff estimation (SCS, 1972; NRCS, 2015). Wang and Tang (2014) used the hypothesis to identify commonalities in empirical functional representations of precipitation partitioning at event, seasonal and annual time scales (including the Budyko equation, the SCS Curve Number method, and other approaches). At event scales, for instance, precipitation is apportioned to soil wetting and runoff. Soil wetting includes both the initial abstraction (before runoff begins) and continuing abstraction. The upper limit of continuing abstraction is constrained by the maximum potential water retention in the landscape (e.g., soil moisture and surface depression storage capacities). Runoff is limited by available water (P minus initial abstraction). The PH dictates that the ratio of continuing abstraction to storage capacity equals the ratio of Q to available water. Some form of the PH appears in multiple contexts in methods applied at the event, monthly, seasonal, and annual time scales. A generalized version is (Wang & Tang, 2014):

$$\frac{X - X_o}{X_p - X_o} = \frac{Y}{Z - X_o} \tag{8.17}$$

Water inputs Z are partitioned into components X, Y. X (e.g., soil wetting in the curve number method) is constrained by its potential value X_p (e.g., storage capacity). The initial abstraction X_o must be supplied by X first. Y (runoff in most models) is constrained by total water availability $Z - X_0$.

Wang et al. (2015) went on to apply the PH to interpret Budyko curves and the L'vovich (1979) formulation of the water balance. The latter is a two-stage competition whereby P is partitioned into soil wetting (W) and fast runoff (Q_f) in the first stage. In the second stage, when initial soil wetting W_o is complete, continued wetting (W-W_o) competes with fast runoff for the available water P—W_0. Applying the MEP principle to the first stage of L'vovich's method (which considers chemical potentials and conductance effects) yields an optimization problem, subject to $P = W + Q_f$ Wang et al. (2015) obtained the optimal solution:

$$\frac{W - W_o}{\alpha_w (W_p - W_0)} = \frac{P - W}{\alpha_f (P - W)_o} \tag{8.18}$$

$$Max\sigma_{stage1}(W, Q_f) = \frac{(W - W_o)^2}{\alpha_w (W_p - W_o)} + \frac{Q_f^2}{\alpha_f (P - W_0)} \tag{8.19}$$

The left side of Eq. 8.19 refers to the maximum entropy production subject to $P = W + Q_f$, and coefficients α_f, α_w represent conductance for runoff and wetting. For the second stage of partitioning, again applying MEP, the optimal solution us

$$\frac{E - E_o}{\alpha_{ec} (E_p - E_0)} = \frac{W - E}{\alpha_s (W - E_o)} \tag{8.20}$$

where α_{ec}, α_s represent conductance for evaporation and slow runoff. An analogous derivation of proportionality from MEP was presented for Budyko models (Wang et al., 2015). Zhao et al. (2016) then used MEP to link the event, monthly, and annual scales. This is another example of coevolution involving water inputs and partitioning and vegetation driven by inputs of solar energy, governed by selection for least action, as represented by maximum entropy production.

These perspectives in the water balance indicate thermodynamic efficiency selection operating in several ways in the evolution and functioning of hydrological systems, ecosystems, and related aspects of landscapes.

8.4 Adaptation

Moving toward a more detailed scale, where storage changes are often critical, we can address whether the concept of adaptation can be applied to Earth surface systems (independently of biological adaptation). One example was already presented in the Soil Hydrology section of Chap. 1 (Sect. 1.3.3), showing adaptive capabilities in natural flow systems.

Here we examine hydrological systems at a scale nested within the water balance approaches discussed above, examining the relationships between storage within a landscape-scale hydrological system and flow leaving the system.

Concentrated, preferential flow occurs at all scales in hydrologic systems. Preferential flows shape, and are shaped by, patterns and structures in soils and landforms. As discussed earlier, preferential flow in surface channel networks and dual-porosity subsurface flow systems are a way of achieving or approaching maximum efficiency, as predicted by several theoretical frameworks, including dissipative systems, constructal theory, network evolution, percolation, and ecohydrological theories. All converge on similar predictions and interpretations of preferential flow, answering *why* these patterns form and persist. But, as hydrological, geomorphological and pedological systems have no intentionality or purposeful agency, and cannot consciously seek improved efficiency, *how* these systems evolve is an open question.

An explanation by Phillips (2022) is based on five factors: gradient and resistance selection, positive feedbacks, intersecting flow paths, threshold limitations, and stability. As outlined in Chaps. 4 and 5, concentrated flows form due to principles of gradient and resistance selection. Second, positive feedback reinforces the concentrated preferential flow paths and their relationship to potential moisture storage zones. Third, flow paths intersect to form networks. However (fourth), expansion of networks and preferential flow paths and networks is limited by thresholds of flow needed for channel, macropore, or conduit growth and maintenance. This produces a "store and pour" flow system capable of retaining water during dry periods and exporting it efficiently during wet periods, providing they develop "spillway" and/or secondary storage mechanisms to accommodate excess water. Finally, store-and-pour systems are typically dynamically stable. Thus, they persist and are selected for. Store-and-pour structures are beneficial for flow system survival, and to support vegetation and ecosystems (i.e., the geophysical and ecological jobs referred to in Sect. 1.3.3). As flow systems, landforms and soils cannot actively pursue goals, and no laws dictate evolution toward such patterns, their emergence and persistence are a matter of selection; survival of the most stable.

Budkyo-type water balance analyses boil down to precipitation and evapotranspiration and water or energy limitations, but at the more detailed scale the essence is "pour" (flow) vs. "store" (moisture storage and delayed flow), limited by internal store vs. pour interactions, and self-limiting or self-reinforcing feedbacks to flow and storage.

A simple qualitative model of the interrelationships between the store and pour components is shown in Fig. 8.3. Flow and storage components may have negative relationships on each other (links a_{12}, a_{21} in Fig. 8.3), when there is a fixed quantity of excess moisture partitioned to one or the other—i.e., stored moisture or slow flow reduces the water in preferential flow paths, and vice-versa. However, flow and storage can also be mutually reinforcing during high inputs, for example via interflows, channel-wetland-floodplain connectivity, and infiltration from flow paths (transmission losses). Storage in the soil matrix and in aquifers may enhance preferential flow

Fig. 8.3 Store and pour interactions

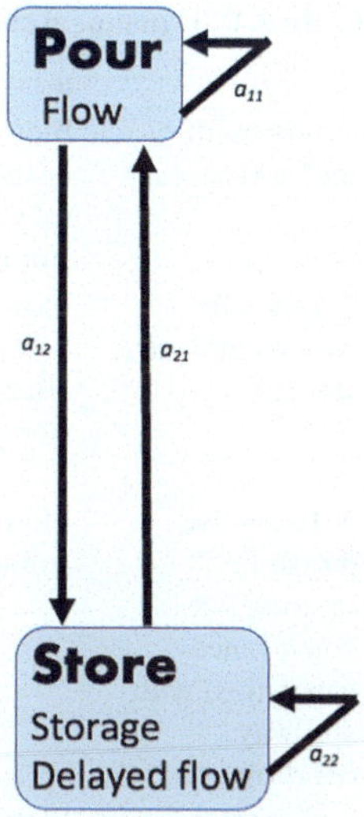

via exfiltration and return flow, and conduit and channel flows may feed subsurface storage.

Self-effects (links a_{11}, a_{22}) are also present. These can be self-limiting, where factors such as soil moisture storage capacity or channel or conduit conveyance capacity limit storage or flow. Self-effects may also be self-reinforcing (positive). Examples include increased flow depths in channels, which inundates roughness elements, increasing conveyance capacity, and increases in hydraulic conductivity by wetting, increasing recharge of storage.

The Routh-Hurwitz criteria for stability requires

$$a_{12}a_{21} - a_{11}a_{22} < 0. \tag{8.21}$$

If pour and store components are mutually limiting (competitive) or mutually reinforcing (a_{12}, a_{21} are both negative or both positive) the system is dynamically unstable unless negative self-limiting effects are dominant. If a_{12}, a_{21} have opposite signs the system is dynamically stable, provided that self-effects are negative (or that negative a_{11} or a_{22} are stronger than positive).

Table 8.1 shows stability analyses of various scenarios. Stability under dry scenarios depends on moisture storage (scenarios 1,2), as limited storage capacity is linked to vegetation stress and mortality and to desiccation. Under excessive moisture inputs (defined as greater than normal storage and flow path conveyance capacities), dynamical instability occurs when erosion by high flows enlarges channels, conduits, etc.—unless the expanding flow paths capture stored water in sufficient quantities (scenarios 5A, 5B). Limited spillway or secondary storage capacity also creates instability (scenario 6). The presence of spillways to convey excess flows creates dynamical stability (scenario 3), as does secondary storage capability (if not overwhelmed; scenario 4).

Table 8.1 Dynamical stability of pour/flow vs. storage relationships in several scenarios

Scenario	a_{12}	a_{21}	a_{11}	a_{22}	*Stability*
1. Dry, with moisture storage	Flow enhances or has no effect on storage	Retention of stored water reduces flow	Negligible or negative	Negative due to increasing tension	S
2. Dry with limited moisture storage	Flow enhances or has no effect on storage	Storage enhances flow due to limited retention, or no effect	May be positive, negative, or negligible	Negligible, or negative due to increasing tension	U
3. Excessive moisture (>storage & conveyance capacities) with spillway overflow	Flow reduces storage	Storage enhances flow by feeding spillways	Negative due to finite conveyance capacity	Negative due to finite storage capacity	S
4. Excessive moisture (>storage & conveyance capacities) with secondary storage	Flow enhances storage by filling depressions, cavities, gravity water	Storage limits flow by activation of secondary storage	Negative due to finite conveyance capacity	Positive, as saturation or filling activates secondary storage	CS[a]
5A. Excessive moisture with erosional enlargement of flow pathways	Flow reduces storage	Storage reduces flow	Positive due to increasing conveyance capacity	Negative due to finite storage capacity	U
5B. Excessive moisture with erosional enlargement of flow pathways	Flow reduces storage	Storage enhances flow due to moisture capture by expanding channels, macropores, or conduits	Positive due to increasing conveyance capacity	Negative due to finite storage capacity	CS[b]
6. Excessive moisture with limited spillway or secondary storage capacity	Positive or negligible	May be positive, negative, or negligible	Negative due to finite conveyance capacity	Positive	U

After Phillips (2022, Table 1)
S stable, *U* unstable, *CS* conditionally stable
[a]Stable if conveyance capacity self-effects stronger than storage ($a_{11} > a_{22s}$); unstable otherwise
[b]Stable if flow-storage feedbacks a_{12}, a_{21}) greater than self-effects, or if $a_{22} > a_{11}$; unstable otherwise

Selection should thus "choose" for storage capacity to retain water during dry periods, and development of spillways and auxiliary storage—that is, storage and flow pathways or compartments that are activated in wet conditions. One example illustrating selection for stability in coevolution of flow systems, soils, vegetation, and landforms is forest soils (e.g., Sidle et al. 2001; Neiber et al., 2006; Nieber and Sidle, 2010; Liu and Lin, 2015; Wilson et al., 2017; Šamonil et al., 2020). Others include shrub-dominated drylands (e.g. Neave and Abrahams, 2002; Wainwright et al., 2002; Abrahams et al., 2003; Parsons et al., 2003; Schymanski et al., 2010); and channel-wetland complexes in low-gradient coastal plain rivers (Phillips, 2022, 2023). Counter-examples where absence of adequate storage or spillway capacity are selected against (that is, the systems are transformed and do not persist) are given in Phillips (2022).

Adding inputs and outputs to the store and pour components within the system results in more detailed determination of stability conditions in various scenarios, as outlined by Phillips (2023) and summarized in Sect. 1.3.3.

8.5 Case Studies

Landscape evolution in specific places involves coevolution of climate, landforms, soils, ecosystems, and hydrology. Because real landscapes involve dense interconnections of biotic and abiotic entities, geophysical and abiotic selection, natural selection, and ecological filtering are all involved. In this section we review three illustrative examples, but potential examples are endless (Phillips, 2021).

8.5.1 White Desert

In Phillips (2025), I used Egypt's White Desert as an example of TREE—Transformative, Recursive, Emergent Evolution. TREE emphasizes that landscape evolution often involves environmental transformations (e.g., regime shifts, metamorphosis), mutual adjustments, and emergent phenomena.

The section of the Egyptian Sahara west of the River Nile is known as the Western Desert and includes a karst massif known for its spectacular landforms and white rock outcrops—the White Desert. The landforms are primarily karst in origin—but karst requires humid conditions, while the White Desert is not just dry, but hyperarid. The region thus represents an example of landscape transformation over long geological timescales that includes strong elements of inheritance and historical contingency.

In the Bahariya-Farafra Oasis areas 16 fields of paleokarst are present (Fig. 8.4). These include karstified late Cretaceous carbonate platforms, where dissolution was active from the late Cretaceous to the late Eocene, guided by structural lineaments comprising E-W- and N-S-trending faults, fractures, and joints—resistance and gradient selection at work. Later came Miocene episodes of uplift and tectonic/climate change favoring karstification. Aeolian and fluvial action affected by local structures modified the karst features (resistance and gradient selection again) (Embabi, 2018; El Aref et al., 2021).

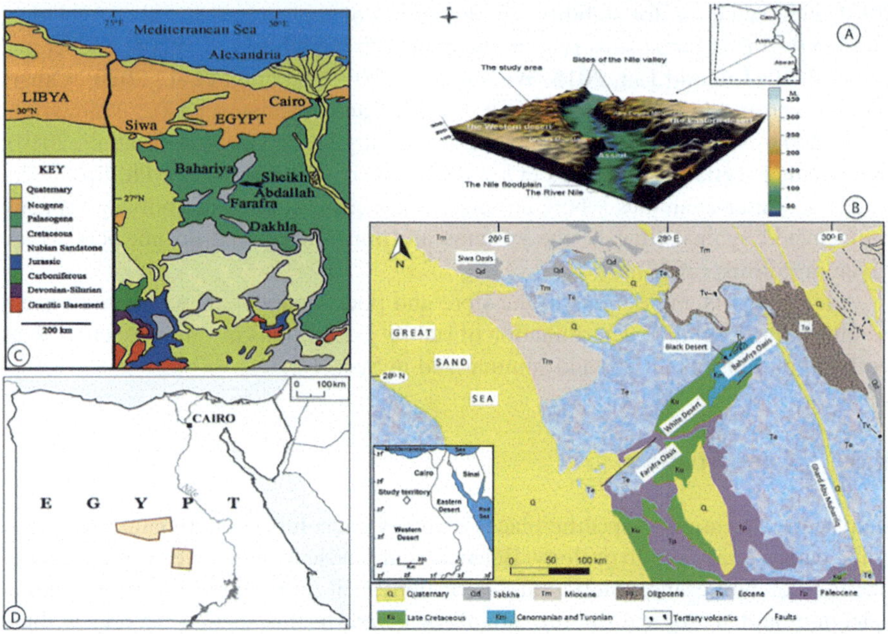

Fig. 8.4 (**a**) Satellite-derived image showing setting of White Desert National Park. (**b**) Geologic Survey of Egypt map of surficial geology. (**c**) Geological map of Egypt. (**d**) Exposures of karstified carbonate rocks exposed in the western desert. (Reprinted under CC-BY-4.0 license from Khalaf (2022). © The Author)

Mushroom rocks are among the most striking landforms. These have 1–5 m stems topped by caps, often bowl-shaped or bulbous (Fig. 8.5). The shapes are a direct result of resistance selection due to lithological variations—resistant caps are mostly hard siliceous limestone, while stems are mainly soft calcareous sandstones. Inselberg-like rock formations shaped by differential dissolutional weathering and sculpted by wind erosion are also common. These are resistant remnants. Resistance selection also produces tower, bridge, and cone karst features. Khalaf (2022) gives a more complete account of White Desert paleokarst features.

White Desert evolution from a humid, karst environment to a hyperarid desert dominated by relict, paleokarst forms occurred between the late Eocene and to the mid-Holocene, when there was a shift from more humid to hyperarid conditions (Kinderman et al., 2006). Intermediate stages in the transformation were influenced by regional and global climate change, tectonic episodes, and biological evolutionary events. Eocene karst vs. hyperarid desert with paleokarst is a highly simplified comparison that overlooks the intervening stages. Nevertheless, the role of resistance and gradient selection is obvious, and natural selection effects over 34 ma can be assumed. Ecological filtering is evident in the effects of climate on biota, and in the key role of highly specialized paleokarst habitats in the modern drylands.

Fig. 8.5 Examples of features in White Desert National Park that originated as karst landforms and subsequently modified by aeolian erosion. (**a**, **b**) Mushroom rocks. (**c**) Wave rocks—wind-abraded features modified from resistant carbonate outcrops. Courtesy of Lynn Phillips. (**d**) Cannonball concretions with karst pavement in foreground. (© 2013 J.D. Phillips)

8.5.2 Fluvial-to-Estuary Transition

As sea-level rises, coastal backwater effects on low-gradient coastal plain rivers propagate upstream, and in these rivers backwater influences extend well upstream of the head of the estuary. Phillips (2024) analyzed the upstream-downstream spatial distribution of these indicators as a space-for-time substitution in 20 coastal plain rivers in Virginia and the Carolinas. Results show that hydraulic effects are the leading edge of sea-level rise (SLR) influences, resulting in higher river stages, increasing floodplain inundation and raising water tables. Reduced hydraulic slopes and flow velocities diminish sediment transport, reducing river sediment input and floodplain deposition. This restricts development of natural levees, thereby reducing bank heights. Those changes increase the frequency and duration of inundation, forming semi-permanently flooded wetlands. Organic decomposition is suppressed by anaerobic conditions, while water ponding allows deposition of transported and suspended organic matter. Thus, organic muck and peat floodplain soils form. The general Holocene valley-filling, along with this muck and peat accumulation, buries alluvial terrace remnants. At the downstream end of the fluvial-estuarine transition zone vegetation changes occur, driven by salinity increases, resulting in swamp conversions to brackish marsh. Backwater encroachment is strongly controlled by channel bed slope, with relatively steeper channels experiencing slower rates of tidal extension. The sequential changes as SLR effects move inland are summarized in Fig. 8.6.

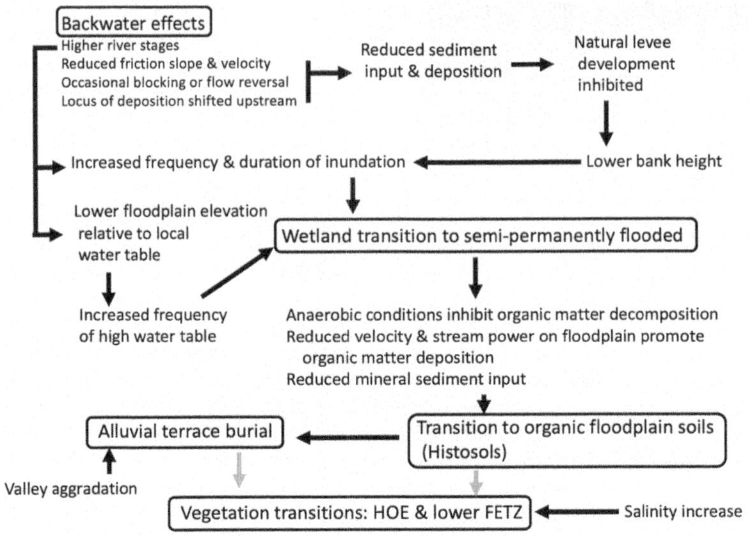

Fig. 8.6 Inferred sequence of transformations in low-gradient coastal plain rivers due to upstream extension of backwater effects. Boxed components are observed transformations in the space-for-time substitution. The other elements shown relate to processes. The black arrows indicate causal relations; gray arrows indicate sequential changes where no apparent direct causal link is known. (Reprinted under CC-BY-4.0 license from Phillips (2024). © The Author)

Resistance selection is key in these transformations, mainly in the form of selective transport and deposition associated with reductions in fluvial sediment transport capacity upstream, and SLR-driven valley filling from downstream.

Rates of backwater encroachment are strongly affected by channel and water surface slopes. In the main channel flow can be represented by the momentum equation in the diffusive wave method:

$$\frac{\partial D_{bf}}{\partial x} - S_c + S_f = 0 \tag{8.22}$$

D_{bf} is bankfull flow depth, x is distance along the channel, and S_c, S_f are the channel and friction slopes, respectively. As river flow encounters the coastal backwater, water surface slope (approximating S_f) flattens out, so that $S_c > S_f$. To maintain momentum balance $\partial D_{bf}/\partial x$ must increase, producing the higher stages and backwater effects.

A first order estimate of upstream encroachment E of SLR effects can be based on a simple geometric argument (that neglects other geomorphic and hydrologic changes), based on the ratio of the amount of SLR divided by channel slope: $E = SLR/S$. Figure 8.7 shows the rate of upstream encroachment for the range of channel slopes encountered in the fluvial-estuary transition zones of the Carolinas

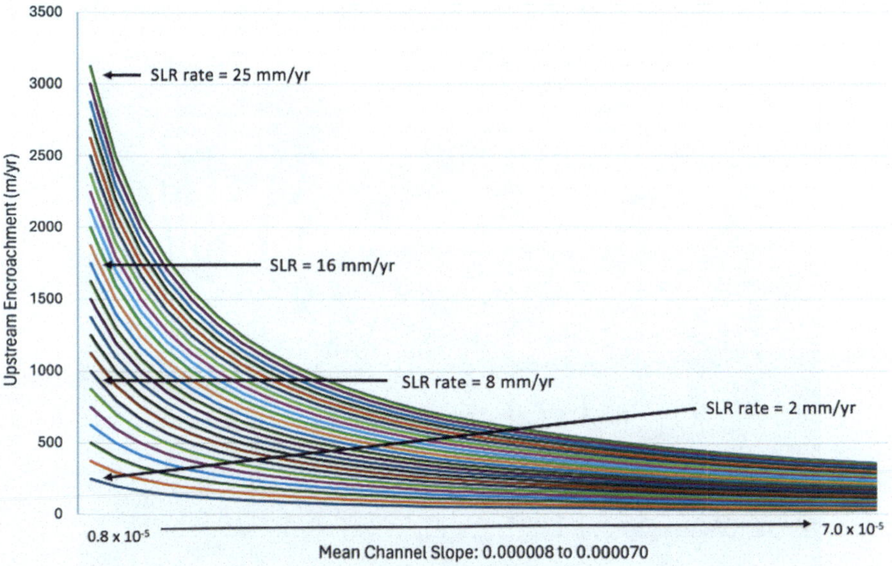

Fig. 8.7 Each curve represents upstream propagation of backwater effects for various rates of sea-level rise for a range of channel slope gradients. (Reprinted under CC-BY-4.0 license from Phillips (2024). © The Author(s))

study rivers for a range of SLR rates from historical twentieth century rates through accelerated recent rates, to rates under various scenarios predicted by NOAA OCM (2022). The wide variation indicates a form of gradient selection in encroachment of SLR effects that operates more rapidly in lower-gradient rivers and tributaries.

8.5.3 Brazilian Cerrado

Central Brazil's Cerrado features a diverse array of landforms and geological settings supporting a geographic mosaic of heterogenous vegetation. The Cerrado is the world's largest Neotropical savanna (Fig. 8.8). This case study is based on the work of Lira-Martins et al. (2022).

Cerrado is a biogeographic region within Brazil that includes grasslands, shrublands, savanna grasslands, and woodland savannas (regional terms are *campos*, *campos sujos*, *cerradão*). The original extent was about 2 million km^2 (much has been altered by conversion to industrial-scale agriculture). Lira-Martins et al. investigated how the complex underlying geology of the region is reflected in geomorphology and soil geography, and in turn edaphic properties and vegetation.

Fig. 8.8 The Cerrado in nineteenth century landscape art. Top: Illustration of vegetation in Minas Gerais. Johann Moritz Rugendas and Alexis Victor Joly, 1835. Bottom: The burning of the campos vegetation in Minas Gerais, Brazil. (Carl Friedrich von Martius, c. 1840. Source: Public domain images discovered through Ballão (2021))

Erosional dissection of the plateau, guided by gradient, resistance, and network selection and preferential flow, produced a variety of landforms which vary in elevation, slope, and drainage context (Fig. 8.9) Differential removal of sedimentary rocks produced a topography that exposes sedimentary layers of varying composition, resistance, and age. The more resistant ridgetops and plateau surfaces are oldest in terms of exposure to weathering, meteorological processes, and biological

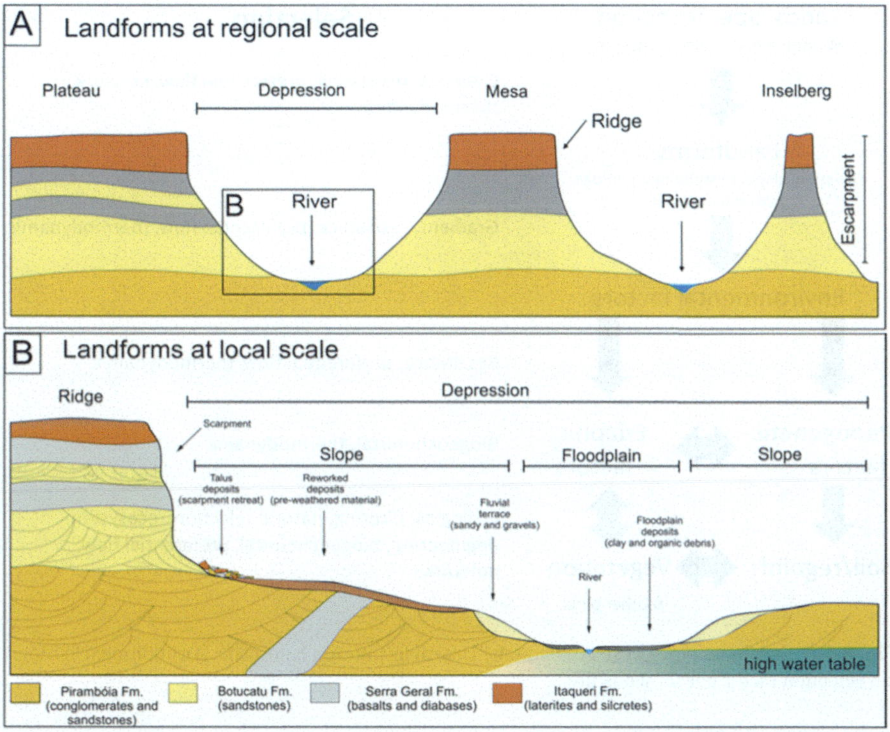

Fig. 8.9 Landforms of the Cerrado. (**a**) Plateaus, mesas, and inselbergs covered by soil or duricrusts are common at regional scales. Landscape dissection exposes geologic materials of variable composition and resistance. (**b**) At local scales, the deposit of material from plateau erosion to lower portions of the slopes represent important geomorphic processes of soil formation. The material deposited next to rivers form soils in fluvial terraces and floodplains. (Reprinted with permission from Lira-Martins et al. (2022).)

effects, while the valley bottoms are the youngest. More morphological features, effects of sediment transport and deposition, and of weathering (note the laterites and silcretes) are revealed at the local scale (Fig. 8.9b). Different sets of pedogenetic factors are associated with these variations, and resistance, thermodynamic, biogeochemical, and ecological selection all operate.

Lira-Martins et al. (2022) also show how soil properties arising from the various selection processes help determine vegetation communities in the Cerrado via ecological filtering. Figure 8.10 below indicates my interpretation of the major forms of selection operating in the system.

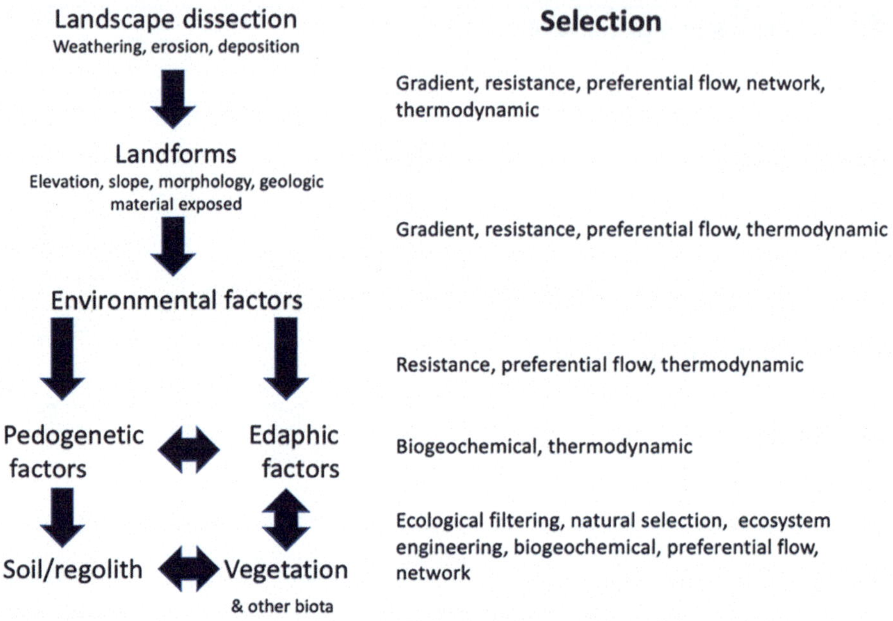

Fig. 8.10 Dominant types of selection in the Brazilian Cerrado landscape. An additional ecological selection factor not shown is fire

8.6 The Elusive Nature of Fitness and Efficiency

If selection is constantly and pervasively operating in ESS, it suggests that landscapes should evolve toward more prevalence of resistant, stable forms, and ever more efficient flux paths and networks. This certainly happens, but the state of maximum efficiency and stability is often not fully attained. Five main factors may work against the attainment of maximum adaptation, fitness, and efficiency: environmental change, the local and probabilistic nature of selection, energy dissipation, canalization, and multiple simultaneous selection. We will examine these in turn.

The time scales involved in development of ESS are often long enough that boundary conditions change, and disturbances occur. Thus, the target of maximum efficiency may be constantly moving, so that before adaptations can be completed the rules and context are modified. This is evident at present with respect to human responses and adaptations to changes in climate and associated shifts in, e.g., the frequency and severity of floods, droughts, storms, and wildfires. Some human societies have adjusted to prevailing late Holocene trends and conditions, but the latter have changed and continue to do so. This is sometimes described as a "new normal," but a constantly shifting baseline may be a better description.

Selection is local. Efficiency selection at the ESS level reflects tendencies that apply in the aggregate, which may or may not be promoted by local selection in individual cases. Flowing runoff, sliding hillslopes, and local breezes, for instance, cannot respond to anything other than the driving and resisting forces in the immediate vicinity. Gradient and resistance selection in response to these forces to maximize efficiency may or may not contribute to, and may even counteract, system-scale efficiency (e.g., Nanson and Huang, 2017; Jerin and Phillips, 2017; Nogueira et al., 2023). For example, the topographic steering of wind in sand dunes described in Sect. 4.1.3 reflects local gradient selection at the scale of 1–100 m^2, which does not enhance overall efficiency of air flow in a dune field or a weather system. Selection is also probabilistic and imperfect, so that the most efficient, resistant, or stable features or configurations do not invariably prevail.

Least action and its related principle of maximum entropy production mean that work is accomplished using the least amount of energy, which means that any excess energy must be dissipated. Such dissipation may modify the most efficient flow paths in ways that counteract maximum efficiency. A good example is turbulent fluid flows, as has been shown by Nanson and Huang (2017) in their work on least action principles in stream channels. Energy dissipation in channels can cause morphological variations and create features that effectively dissipate available energy but work against maximum flow and transport rates (Fig. 8.11).

Canalization may also impair steady progress toward maximum efficiency. Local selection may form flux paths that differ from those promoting efficiency at

Fig. 8.11 Bedrock steps, chutes, and pools effectively dissipate energy in this Montana stream, but work against maximum flow and transport rates. (Reprinted with permission. © 2021, Chris Van Dyke. All rights reserved)

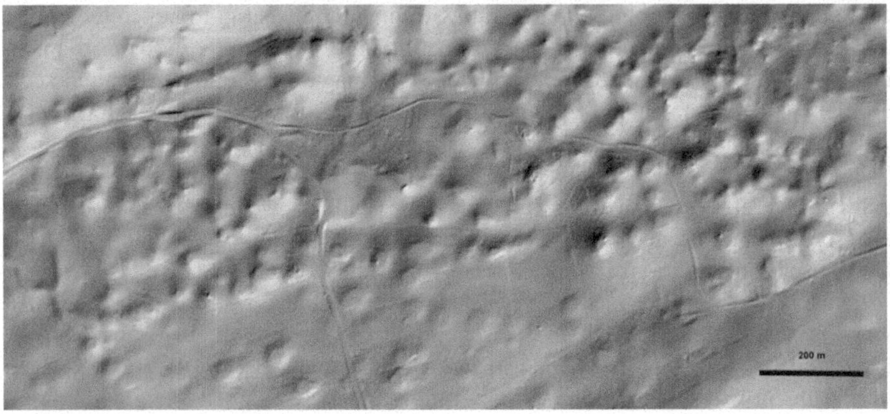

Fig. 8.12 Shaded relief of an area of Washington County, Virginia, showing numerous dolines. Center of image is at 36.659°N, 81.922°W

the broader scale. Once formed, potential future paths are affected. For instance, after they develop the karst sinkholes (dolines) shown in Fig. 8.12 direct water and solute flow, regardless of aggregate efficiency of the hillslopes or drainage basin.

Multiple simultaneous selection refers to the fact that in any landscape selection happens simultaneously on many different materials, forms, processes, flow paths, interrelationships (and organisms) at any given moment. The selection criteria differ, and increasing efficiency or stability for one may not enhance that of another and may even offset or prevent it. Paths taken by burrowing fauna to maximize their own efficiency, for instance, may not be the most efficient routes for water flows that later follow them. Contrasting trends also occur in biogeochemical selection, where in some cases rooting strategies to optimize the carbon budget differ from those optimizing water use. Dam-building by humans or beavers may make for better resource procurement, safety, and convenience for the dam builders, but does not improve the efficiency of stream flow. Such offsetting effects are particularly evident in force-resistance relationships. Resistance selection leads to preferential preservation of more resistant materials and features, but this is relative to force. If more resistant rock, for instance, coincides with the most efficient flow path for abrasion-driving winds with higher velocities, the more resistant rock may not be preserved.

In summary, while efficiency selection consistently operates in landscapes, a progression toward greater efficiency is not always observed. This is due to the probabilistic and local nature of selection, and changes in dominant controls of landscape evolution, and therefore of selection criteria, over time. Energy dissipation in one process may decrease efficiency in another. Selection operates on multiple components simultaneously, with effects that may counteract each other.

8.7 Common Themes

Whether dealing with specific regions or landscapes or the broader scope of atmosphere, hydrosphere, and biosphere interactions, several common themes emerge as we consider selection in landscape and ESS evolution.

One is *coevolution*. Just as species do not evolve in isolation from each other and their environment, climate, hydrology, landforms, soils, and ecosystems coevolve. It is difficult, if not impossible, to understand evolution of the water balance, or of the global atmosphere, biosphere, or hydrosphere without considering (at least) all three components. Local or regional scale landscape evolution involves coevolution of landforms, soils, ecosystems, and hydrological systems, and in many cases feedbacks to climate and microclimate must also be considered. A related theme is that in ESS multiple forms or selection are often operating simultaneously. Sometimes these are complementary (as in the PT-based studies mentioned above), but in other cases they may be independent or counteracting.

TREE is another recurring theme. Landscape evolution is not always transformative in the sense of leading to new system states, but often is, at geological time scales (e.g., the White Desert) and historical/contemporary time scales (e.g., fluvial-estuarine transition zones). Recursivity is inevitably present, as for example between vegetation, climate, and hydrology, or between plants, soils, landforms, and water flows. Emergence is essentially an outcome of selection—rather than a guiding hand or law, evolution is nudged by selection, and landscape characteristics emerge accordingly.

The intertwined development of soil, hydrology, landforms, and biota can be linked by the fact that water, solutes, and roots must all pass through the porous medium of soils, with energy and nutrient fluxes following along. This is what enables constructs such as percolation theory and other network-based approaches to identify common selection processes in terms of minimizing resistance.

Mass (especially water) and energy partitioning are another common thread. The store-and-pour and Budyko and water balance constructs are all based on how matter, energy, and resources are partitioned, and selection effects on this partitioning are the key to adaptation of ESS.

In Chap. 9 we look more closely at emergent phenomena and how selection can explain various proposed or apparent goal functions in ESS evolution.

References

Abrahams, A. D., Parsons, A. J., & Wainwright, J. (2003). Deposition of rainwater under creosotebush. *Hydrological Processes, 17*, 2555–2566.

Ballão, A. S. (2021). The Brazilian *Campos* in nineteenth-century landscape art. *Arcadia, 15*. https://www.environmentandsociety.org/arcadia/brazilian-campos-nineteenth-century-landscape-art

Budyko, M. I. (1974). *Climate and life*. Academic.

Egli, M., Hunt, A. G., Dahms, D., et al. (2018). Prediction of soil formation as a function of age using the percolation theory approach. *Frontiers in Environmental Science, 6*, 108.

El Aref, M. M., Salama, A., & Hammed, M. (2021). Morphotectonic evolution of Qaret El Sheikh Abdallah depressions and denuded paleokarst in the White Desert, El Bahariya-Farafra Karst territory, Egypt. *Egyptian Journal of Geology, 65*, 27–53.
Embabi, N. S. (2018). *Landscapes and landforms of Egypt: Landforms and evolution* (World Geomorphological Landscapes Series). Springer.
Hunt, A. G. (2016). Spatio-temporal scaling of vegetation growth and soil formation from percolation theory. *Vadose Zone Journal, 15*, 1–15. https://doi.org/10.2136/vzj2015.01.0013
Hunt, A. G. (2017). Spatio-temporal scaling of vegetation growth and soil formation: Explicit predictions. *Vadose Zone Journal, 16*, 1–12.
Hunt, A. G. (2024). The physics and the biology of the water balance: A personal journey through the critical zone into the water balance. *Hydrological Processes, 38*, e15209.
Hunt, A. G., Faybishenko, B., & Ghanbarian, B. (2021). Non-linear hydrologic organization. *Nonlinear Processes in Geophysics, 28*, 599–614.
Hunt, A. G., Ghabarian, B., & Faybishenko, B. (2023a). A model of temporal and spatial river network evolution with climatic inputs. *Frontiers in Water, 5*, 1174570.
Hunt, A. G., Faybishenko, B. A., & Powell, T. L. (2020). A new phenomenological model to describe root-soil interactions based on percolation theory. *Ecological Modelling, 433*, 109205. https://doi.org/10.1016/j.ecolmodel.2020.109205
Hunt, A. G., Sahimi, M., & Ghanbarian, B. (2023b). Predicting streamflow elasticity based on percolation theory and ecological optimality. *AGU Advances, 44*, e2022AV000867.
Hunt, A. G., Sahimi, M., Ghanbarian, B., & Poveda, G. (2024). Predicting ecosystem net primary productivity by percolation theory and optimality principle. *Water Resources Research, 60*, e2023WR036340.
Hunt, A. G., Skinner, T. E., Ewing, R. P., & Ghanbarian-Alavijeh, B. (2011). Dispersion of solutes in porous media. *European Physics Journal B, 80*, 411–432.
Jerin, T., & Phillips, J. D. (2017). Local efficiency in fluvial systems: Lessons from Icicle Bend. *Geomorphology, 282*, 119–130.
Khalaf, E. A. (2022). Karst Heritage as a tourist attraction: A case study in the White Desert National Park, Western Desert, Egypt. *Geoheritage, 14*, 94.
Kinderman, K., Bubenzer, O., Nussbaum, S., et al. (2006). Palaeoenvironment and Holocene land use of Djara, Western Desert of Egypt. *Quaternary Science Reviews, 25*, 1619–1637.
Lira-Martins, D., Nascimento, D. L., Abrahão, A., et al. (2022). Soil properties and geomorphic processes influence vegetation composition, structure, and function in the Brazilian Cerrado. *Plant and Soil, 476*, 549–588. https://doi.org/10.1007/s11104-022-05517-y
Liu, H., & Lin, H. (2015). Frequency and control of subsurface preferential flow: From pedon to catchment scales. *Soil Science Society of America Journal, 79*, 362–377.
Loague, K., Corwin, D. 2006. Chapter 25: Scale issues. In: J.W. Delleur, editor, The handbook of groundwater engineering. 2 CRC Press, ., 25.1- to 25.21.
L'vovich, M. I. (1979). *World water resources and their future.*, 415 pp. AGU.
Nanson, G. C., & Huang, H. Q. (2017). Self-adjustment in rivers: Evidence for least action as the primary control of alluvial-channel form and process. *Earth Surface Processes and Landforms, 42*, 575–594.
Neave, M., & Abrahams, A. D. (2002). Vegetation influences on water yields from grassland and shrubland ecosystems in the Chihuahuan Desert. *Earth Surface Processes and Landforms, 27*, 1011–1020.
Nieber, J. L., Steenhuis, T. S., Walter, T., & Bakker, M. (2006). Enhancement of seepage and lateral preferential flow by biopores on hillslopes. *Biologia, 61*, 225–228.
Nieber, J. L., & Sidle, R. C. (2010). How do disconnected macropores in sloping soils facilitate preferential flow? *Hydrological Processes, 24*, 1582–1594.
Nijzink, R. C., & Schymanski, S. J. (2022). Vegetation optimality explains the convergence of catchments on the Budyko curve. *Hydrology and Earth System Sciences, 26*, 6289–6309.

NOAA OCM (U.S. National Oceanographic and Atmospheric Administration, Office of Coastal Management). 2022. Sea level rise viewer. Available online: https://coast.noaa.gov/slr/. Accessed on 31 August 2024.

Nogueira, A. M., Caldeira, U. R., et al. (2023). Sedimentary records in dry karstic valleys: The case of Mambaí, Central Brazil. *Journal of South American Earth Sciences, 128*, 104338.

NRCS (Natural Resources Conservation Service, U.S. Department of Agriculture). (2015). National Engineering Handbook: Part 630 - Hydrology.

Parsons, A. J., Wainwright, J., Schlesinger, W. H., & Abrahams, A. D. (2003). The role of overland flow in sediment and nitrogen budgets of mesquite dunefields, southern New Mexico. *Journal of Arid Environments, 53*, 67–71.

Phillips, J. D. (2021). *Landscape evolution. Landforms, ecosystems, and soils*. Elsevier.

Phillips, J. D. (2022). Store and pour: The evolution of flow systems in landscapes. *Catena, 216*, 106357.

Phillips, J. D. (2023). Contingent partitioning and adaptation in hydrological systems. *Ecohydrology, 16*, e2567. https://doi.org/10.1002/eco.2567

Phillips, J. D. (2024). Sequential changes in coastal plain rivers affected by rising sea-level. *Hydrology, 11*, 124. https://doi.org/10.3390/hydrology11080124

Phillips, J. D. (2025). *Mysterious ways. Contingency, emergence, and selection in nature*. Oxford University Press.

Qui, N., Xi, C., Hu, Q., et al. (2018). Hydro-stochastic interpolation coupling with the Budyko approach for prediction of mean annual runoff. *Hydrology and Earth System Sciences, 22*, 2891–2901. https://doi.org/10.5194/hess-22-2891-2018

Reaver, N. G. F., Kaplan, D. A., Klammler, H., & Jawitz, J. W. (2022). Theoretical and empirical evidence against the Budyko catchment trajectory conjecture. *Hydrology and Earth System Sciences, 26*, 1507–1525.

Renschler, C. S. (2000). *Strategies for implementing natural resource management tool—A geographical information science perspective on water and sediment balance assessment at different scales*. PhD dissertation, University of Bonn.

Rosenzweig, M. L. (1968). Net primary productivity of terrestrial communities: Prediction from climatological data. *The American Naturalist, 102*, 67–74. https://doi.org/10.1086/282523

Šamonil, P., Phillips, J. D., Daněk, P., & at al. (2020). Soil, regolith, and weathered rock: Theoretical concepts and evolution in old-growth temperate forests, Central Europe. *Geoderma, 368*, 114261.

Schymanski, S. J., Kleidon, A., Steiglitz, M., & Narula, J. (2010). Maximum entropy production allows a simple representation of heterogeneity in semiarid ecosystems. *Philosophical Transactions of the Royal Society B, 365*, 1449–1455.

Sidle, R. C., Noguchi, S., Tsuboyama, Y., & Laursen, K. (2001). A conceptual model of preferential flow systems in forested hillslopes: Evidence of self-organization. *Hydrological Processes, 15*, 1675–1692.

SCS (Soil Conservation Service, U.S. Department of Agriculture). (1972). *National engineering handbook, Section 4, Hydrology*. U.S. Government Printing Office.

Sposito, G. (2017). Understanding the Budyko equation. *Water, 9*, 236.

Wainwright, J., Parsons, A. J., Schlesinger, W. H., & Abrahams, A. D. (2002). Hydrology-vegetation interactions in areas of discontinuous flow on a semi-arid bajada, southern New Mexico. *Journal of Arid Environments, 51*, 319–338.

Wang, D., & Tang, Y. (2014). A one-parameter Budyko model for water balance captures emergent behavior in Darwinian hydrologic models. *Geophysical Research Letters, 41*, 4569–4577. https://doi.org/10.1002/2014GL060509

Wang, D., Zhao, J., Tang, Y., & Sivapalan, M. (2015). A thermodynamic interpretation of Budyko and L'vovich formulations of annual water balance: Proportionality hypothesis and maximum entropy production. *Water Resources Research, 51*, 3007–3016.

Wilson, G. V., Nieber, J. L., Fox, G. A., et al. (2017). Hydrologic connectivity and threshold behavior of hillslopes with fragipans and soil pipe networks. *Hydrological Processes, 31*, 2477–2496.
Yu, F., & Hunt, A. G. (2017). Predicting soil formation on the basis of transport-limited chemical weathering. *Geomorphology, 301*, 21–27. https://doi.org/10.1016/j.geomorph.2017.10.027
Yu, F., Faybishenko, B. A., Hunt, A. G., & Ghanbarian, B. (2017). A simple model of the variability of soil depths. *Water, 9*, 460. https://doi.org/10.3390/w9070460
Zhao, J., Wang, D., Yang, W., & Sivapalan, M. (2016). Unifying catchment water balance models for different time scales through the maximum entropy production principle. *Water Resources Research, 52*, 7503–7512.

Goals and Emergence

9

Abstract

What, if anything, guides the evolution of Earth surface systems (ESS)? Goal functions and attractors are attempts to explain or predict where landscape evolution is headed, and why. Goal functions and attractors may be roughly grouped into three categories—single path, single outcome; alternative stable states; and multiple path, multiple outcomes. The latter includes systems characterized by dynamical instability and deterministic chaos, where small disturbances or variations in initial conditions become magnified (amplification selection). A classic example—the Lorenz model of atmospheric convection—is discussed. Traditional theories of succession or progressive development to a climax form are single path/outcome attractors common in geology, geomorphology, pedology, and ecology. Various forms of selection play a key role in each of these. Steady-state equilibrium and self-organized criticality concepts are also single path/outcome constructs, sometimes associated with purported goal functions, but that actually represent emergent selection. The example of river longitudinal profiles is presented. Power law distributions are often presented as attractor states, or as evidence of attractors. However, power law distributions alone are problematic evidence, as many phenomena can produce these. Alternative stable state concepts involve multiple (usually two) stable attractor states separated by instability, typically in threshold-dominated systems. Goal functions that are often described as ESS seeking a particular state can be explained via emergent selection. As this does not imply any intentionality of ESS, the emergent selection explanation is preferred over goal functions by Ockham's Razor.

J. D. Phillips, *Abiotic Selection in Earth Surface Systems*, Geophysics and Environmental Physics, https://doi.org/10.1007/978-3-031-85862-8_9

9.1 Introduction: Attractors and Goal Functions

What, if anything, guides the evolution of Earth surface systems (ESS)? Even for those who believe in a guiding hand, there remains the question of *how* the guider does the guiding. Goal functions and attractors are attempts to explain or predict where landscape evolution is headed, and why. Goal function as used here is an umbrella term for criteria that have been proposed to govern the operation and development of ESS. In Earth and environmental sciences, normative properties or behavior are often associated with balance-of-nature constructs, though this is more often implicit than explicit.

An attractor, in mathematical dynamical systems theory, is a set of numerical values towards which a system tends to evolve, for a range of starting values. More generally, an attractor is a state or condition towards which the ESS tends to evolve. Climax vegetation communities, mature climax soils, peneplains, steady-state equilibria, and self-organized criticality are all examples of attractors that have been proposed for landscape systems. Goal functions include several optimal and extremal principles postulating that ESS seek to maximize, minimize, or optimize something—we have already encountered many of these in earlier chapters.

A seemingly minor issue, once an attractor or goal function is identified, is its interpretation. Traditionally this has been expressed in anthropomorphic language, as (for instance) the ocean-atmosphere system seeking to redistribute heat or a river adjusting its slope to transport the available sediment load, or in neutral terms, simply stating that ESS tend to (e.g.) maximize entropy production or achieve a steady-state relationship between inputs and outputs. The distinction emphasized here is that most attractors and goal functions are advantageous for the survival or efficiency of the system and are therefore selected for. Progress toward a goal or attractor is therefore an emergent phenomenon rather than a preordained outcome or a normative property.

9.2 Types of Attractors

Goal functions and attractors may be roughly grouped into three categories—single path, single outcome; alternative stable states; and multiple path, multiple outcomes.

Single path, single outcome concepts postulate development toward a predefined, often normative, outcome. Examples include erosional landscape progression toward a peneplain or regional erosion surface; progressive pedogenetic development toward a mature, climax, zonal soil; ecological succession toward a stable climax community; and progression toward a steady-state equilibrium or a self-organized critical state. These frameworks allow for some variability in evolutionary trajectories and for outcomes that are similar but not necessarily identical, but both the pathways and outcomes are within fixed, well-defined ranges.

Alternative stable state constructs allow for multiple possible stable states. Some are contingent on specific triggers or switches, such as rising or falling base level, presence or absence of fire, and thresholds such as sea ice cover or carbon sequestration rates. Others are conceived as regular or sporadic alternation between states, such as those associated with ENSO, glacial-interglacial, or seasonal monsoon oscillations. Most alternative stable state models are binary, with the stable states often oppositional and in-between states unstable. Examples are mostly ecological, but ENSO phenomena and aggradation-degradation phases in fluvial systems have also been interpreted in this context.

Situations where more than two outcomes and multiple evolutionary pathways are possible from identical (or nearly so) starting conditions are relatively common in ESS—multi-path, multi-outcome concepts describe these, which often produce complex geographical patterns and extensive spatial variability over small areas. These are sometimes called nonequilibrium models and are often associated with deterministic chaos and dynamical instability. Though the variability and complexity in ESS often associated with this type of development may appear pseudorandom, ESS evolution is constrained by invariant laws and by the available mass, energy, and geographical space. In this view, ESS evolution can be like chess, with a single starting condition, limited space, strict rules governing play—and an innumerable possible number of move sequences and outcomes.

9.2.1 Instability, Chaos, and Amplification Selection

Nonequilbrium concepts are connected to attractors in dynamical systems by the notion of unstable attractors. Dynamical systems are often represented by a multidimensional phase space defined by two or more key variables of the system. In numerical simulations, a region of phase space to which evolutionary trajectories converge is an attractor. Attractors may be very small regions—even single points—or periodic. More complex attractors are called chaotic or strange attractors.

The canonical example of a chaotic attractor comes from atmospheric science. Edward Lorenz (1963) formulated a system of nonlinear ordinary differential equations to model atmospheric convection, where x, y, z are the state variables of the system, and σ, ρ, and β, respectively, are the Prandtl number (ratio of kinematic viscosity to thermal diffusivity), Rayleigh number (a product of the Prandtl number and the ratio of buoyancy to viscous forces), and a physical proportion:

$$\frac{dx}{dt} = \sigma(y - x) \tag{9.1}$$

$$\frac{dy}{dt} = x(\rho - z) - y \tag{9.2}$$

$$\frac{dz}{dt} = xy - \beta z \tag{9.3}$$

State variable x is the rate of convective overturning, y is the horizontal and z the vertical temperature variation.

The Lorenz system is famous for its chaotic behavior, where miniscule changes in initial parameter values out to many decimal places produce different results. However, rather than filling the x, y, z phase space, as a truly random system would, the trajectories fall on a strange attractor resembling a butterfly (Fig. 9.1).

Chaotic behavior has been demonstrated in a variety of geophysical phenomena, such as atmospheric dynamics, turbulent fluid flows, precipitation, stream flow, earthquake dynamics, and topographic evolution (e.g. Turcotte, 1997; Phillips, 1999, 2006; Sivakumar, 2017). However, few studies are based on or supported by empirical evidence (as opposed to simulation models), and directly linked to landscape evolution rather than shorter-term phenomena. Deterministic chaos has been demonstrated multiple times for hydrological phenomena such as stream flow and rainfall, for instance (Sivakumar, 2017), but that does not necessarily translate directly to a chaotic attractor in the evolution of fluvial networks or groundwater flow systems.

Some studies show dynamical instability and chaos directly in landscape evolution phenomena (e.g., soil formation; Phillips, 2017), but these do not yield the mathematically precise and fascinating visual patterns of the Lorenz attractor or other phase space attractors. Chaos and other forms of complex nonlinear dynamics do not identify an end-state for system evolution; rather they define a multidimensional space within which evolution occurs. To return to the chess metaphor, they

Fig. 9.1 The Lorenz attractor. All trajectories are confined to the attractor, but none repeat

can allow delineation of the chess board and identification of the rules, but not specific sequences of moves or game outcomes.

Where chaos and dynamical instability occur, they imply *amplification selection.* Certain features or changes grow disproportionately large relative to their initial amplitude or causal disturbance. Where such changes last long enough to be reflected in ESS development, they fit the VSR logic of variation, selection, and retention. Amplification selection differs from the other forms discussed here in that it does not necessarily promote the efficiency, stability, or resistance of the system. The chaotic amplification of the effects of a butterfly's wing flapping in Brazil to produce a tornado in Texas (to use a metaphor made famous by) does not improve the fitness or adaptation of the atmosphere in any directly discernible way. However, the mechanisms by which amplification occurs often directly involve other forms of selection. For example, dynamical instability of wetting fronts results directly from gradient and resistance selection.

9.2.2 Selection and Climax

Single-path, single-outcome frameworks for ESS development are now largely agreed to be too limiting to describe the variety of evolutionary trajectories, and are generally seen as possibilities or benchmarks rather than normative trends. Nonetheless, the rise and influence of these conceptual frameworks reflects the fact that they do generally reflect a plausible, coherent logic supported by observational evidence. That is, soils (for instance) do not always, inevitably, progressively approach a mature climax zonal soil, and hillslopes (for instance) do not always maintain a critical inclination near the angle of repose. But these things do indeed happen and can be elucidated rationally.

9.2.2.1 Succession to Climax

The best-known theory of ecosystem development, based on plant communities, is that of Clements (1916). Communities are superorganisms, Clements argued, that progressively develop via a sequence of stages toward a stable endpoint called the climax community. Climax communities are defined by climate and are maintained indefinitely in the absence of serious disturbance or environmental change. Each stage in the sequence facilitates its eventual replacement by the next stage up to the climax. Cowles (1899) also conceived succession in terms of gradual change in plant communities whereby vegetation of one stage facilitates the transition to the next. In Cowles view, however, succession is not necessarily a linear sequence, but a complex, dynamic process that can be regressive and does not necessarily lead to a single climax endpoint. As an alternative to Clements' monoclimax, Tansley (1935) proposed the ecosystem concept to stress importance of abiotic factors other than climate (van der Walk, 2014). Tansley did accept succession toward a stable equilibrium but argued that the climax was not necessarily defined wholly or chiefly by climate, emphasizing the role of soils.

Histories of ecology typically place the individualistic concept of plant associations of Gleason (1939) as the main alternative to Clementsian succession. Plant communities are a function of which propagules (seeds, rootstock, etc.) are available and of ecological filtering in Gleason's framework. Gleason's (1939) concepts are the basis for theories of ecological assembly. There are several of these, but all are based on the Gleasonian notion that chance in the arrival of potential organisms is important (particularly the first, or pioneer, arrivals). Some so-called "neutral" assembly theories do not include niches or ecological filtering—like moisture storage in the Budyko framework, their reality is not denied, but neutral models assert that community composition can be explained without them. Other assembly theories, closer to Gleason's framework, acknowledge a key role for environmental characteristics to allow or facilitate some species and eliminate or inhibit others (see Vellend, 2016 for a review).

Beginning in the 1970s succession theories allowing for more complex and variable successional patterns were presented (Gersmehl, 1976; Cattelino et al., 1979; Westoby et al., 1989). Reviews and syntheses of succession theories by Mori (2011) and Pulsford et al. (2014) show that succession, or ecological change more generally, may be linear (or cyclical, when disturbance resets the system), convergent, divergent, or characterized by various possible transitions among system states. Though they were motivated by and are often associated with alternative stable state concepts, state-and-transition models (STM) are a general tool for describing and predicting ecological state changes that can accommodate virtually any pathway (Briske et al., 2005; Bestelmeyer et al., 2009). STMs, though not always labelled as such, have also been used in the geosciences (Phillips & Van Dyke, 2017). Both linear succession models and STMs can be symbolically labelled, with system state at time t indicated by $S(t)$, as

$$S(t) = f(S_{t-1}) \tag{9.4}$$

but with linear succession there is a fixed progression from $S(1)$, $S(2)$, ... $S(n)$, whereas in STMs the steps or stages may be more varied and are often reversible.

Niche differentiation vs. niche convergence vs. niche construction is an important distinction in evolutionary ecology. Niche differentiation predicts competing species will evolve to use different resources and interact less. Niche-convergence, linked to neutral theories of community assembly, predicts that species evolve to use similar resources and increase their interactions. Niche construction occurs when organisms modify the abiotic environment to create or expand niches for themselves and/or other organisms. This occurs via ecosystem engineering, with niche construction generally referring to the longer-term effects of ecosystem engineering on natural selection. The interactions among niche partitioning, convergence, and construction, evolution, and ecosystem functions for plants are discussed by Thorpe et al. (2011).

Table 9.1 summarizes the role of selection in succession and related theories. Most depend on ecological filtering by abiotic constraints and resources (though not exclusively abiotic). The stable climax or steady-state equilibrium communities postulated to develop given enough time and lack of disturbance imply resistance

Table 9.1 The role of selection in theories of development of ecological communities

Concept or theory	Role of selection
Clementsian succession-to-climax	Ecological filtering to provide climate boundary conditions Resistance selection for stable climax state
Cowles' contingent succession	Ecological filtering to provide climate boundary conditions Contingent ecological filtering→variable pathways Resistance selection for stable climax state
Tansley's ecosystem concept	Ecological filtering by abiotic factors Connection or network selection? Biogeochemical selection
Gleason's plant sociology Multiple path succession	Ecological filtering by abiotic factors & biotic interactions Contingent ecological filtering → variable pathways
State-and-transition models	Ecological filtering by abiotic factors, biotic interactions, disturbance histories & regimes, & management
Niche differentiation	Ecological filtering by abiotic factors & biotic interactions promotes resource partitioning
Niche convergence	Biogeochemical selection Network selection
Niche construction Ecosystem engineering	Ecological filtering by biotic-abiotic interactions

selection for stability. Tansley's ecosystem-based notions of succession imply, or are at least consistent with, biogeochemical selection, as is niche convergence. Ecosystems and niche convergence also imply network selection via favoring the formation of interconnections.

Succession and ecosystem development concepts that allow for multiple possible pathways and outcomes depend on what could be called contingent ecological filtering. That is, spatial variations in habitat characteristics, along with environmental changes and disturbances, can establish or divert various successional pathways.

9.2.2.2 Progressive Pedogenesis and Zonal Soils

Soils are a product of the combined, interacting effects of geology, biota, climate, hydrology, topography, and time (the age of a soil or the surface on which it is formed, or the time span over which pedogenesis occurs). This is the basis of the functional-factorial approach to pedology, soil geography, and soil survey and mapping, exemplified by the well-known "clorpt" model:

$$S = f\left(cl,o,r,p,t\right)\ldots \tag{9.5}$$

S represents the soil type or a soil property at a location, and the other variables represent, respectively, *cl*imate, *o*rganisms, *r*elief (topography), *p*arent material, and *t*ime. The trailing dots represent factors that are important in some settings but not incorporated in the "clorpt" factors.

This approach was pioneered by Dokuchaev (1883) and Jenny (1941) introduced the clorpt equation. It is not linked by logical necessity to any specific concept of soil evolution. However, it is often associated with single-path, single-outcome progression to climax or zonal soils. This is because Dokuchaev's seminal work was embedded in the idea of climax soils determined mainly by climate, and because the factorial approach was accepted and spread by scientists such as Marbut (1935), who had a view of soil formation analogous to Clementsian succession. That is most soils, if relatively undisturbed, will evolve to a zonal, climax form.

Marbut introduced the concept of a "normal" soil, defined as freely drained mature soils with well-expressed horizons characteristic of general environmental settings. As Marbut was chief of U.S. Soil Survey operations, U.S. soil classification was codified based on three major categories: zonal (normal or climax soils), azonal soils (young, immature soils presumably developing toward their zonal equivalent, and intrazonal soils, whose profiles were dominated by local or idiosyncratic factors of soil formation. The pathway from parent material to azonal soils to zonal soils was implicit.

A model of soil formation based on the net effects of matter and energy additions to the soil, losses and removals from the soil, translocations within the soil, and transformations was developed by Simonson (1959, 1978). Simonson viewed soils as continuously evolving, with the state of the soil at a given time a function of the overall effects of additions, removals, translocations, and transformations of matter and energy, rather than the position along a presumed trajectory toward a zonal or climax soil.

Theories of pedogenesis have generally assumed, explicitly or implicitly, that progressive processes and pathways dominate. During progressive pedogenesis, soils become generally thicker, more vertically differentiated (horizonation), and more strongly developed. An alternative was proposed by Johnson and Watson-Stegner (1987), who presented a soil evolution model based on continuous evolution, and the possibility for both or either progressive or regressive pedogenesis to occur and dominate soil development. They expanded Simonson's (1978) list of processes, linking pedogenetic processes directly to specific regressive and progressive pathways. In this Johnson and Watson-Stegner (1987) extended concepts of Nikiforoff (1949), who was one of the first in Anglophone soil science to envision soil *evolution,* as opposed to linear development toward a predetermined end state.

Johnson and Watson-Stegner's soil evolution model and evidence of dynamical chaos and instability in soils led Phillips (1993a, b, c, 1998) to propose that soil development is inherently nonlinear and can exhibit deterministic uncertainty and sensitivity to minor variations in initial conditions and small disturbances. For soil evolution, this implies the possibility of multiple pathways and outcomes. Thus, there may be divergent rather than convergent pedogenesis at the landscape scale. The various ways that divergent pedogenesis and other forms of complexity may arise in soil systems are reviewed in Phillips (2017).

The effect of the soil-forming factors (environmental controls or state factors) in determining soil properties could be described as selection via pedological filtering, a direct analogy of ecological filtering. Parent material, climate, etc. encourage and facilitate development of some soil properties and soil types, while impeding or eliminating others. This form of selection is at least implicit in all the conceptual models. The mature zonal climax soil concept also implies resistance selection, in that the climax soil is expected to persist indefinitely in the absence of major disturbance (and mature soils do often exhibit highly resistant features such as argillic and oxic horizons and quartz sand residuals). The process-based models of Simonson and Johnson and Watson-Stegner do not explicitly address selection. But the various inputs, outputs, translocations and transformations involved include many physical, chemical, biophysical, and biochemical processes to which gradient, resistance, and thermodynamic selection must apply. Dynamical instability and chaos, when and where it occurs, further implies amplification selection.

9.2.2.3 Landforms and Topography

Cyclical theories of landscape evolution postulate single-path development toward a final state, with landscape rejuvenation (uplift) reinitiating the sequence. William Morris Davis' (1899, 1909) cycle of erosion (also known as the geographical or geomorphic cycle) is the best known and most influential of these. For landscapes dominated by fluvial denudation, the cycle begins with an episode of uplift. Streams dissect the uplifted terrain as denudation begins. Denudation proceeds through stages of youth, maturity, and old age, during which overall relief decreases, valleys become wider and less steep, channels meander more strongly, and slopes are generally smoothed. The final stage—the peneplain—is an erosional surface graded to base level. Since the theory involves a cycle, peneplains are eventually uplifted, stimulating a new cycle. The erosion-to-peneplain is broadly analogous to succession-to-climax or soil development to climax, except that it explicitly assumes renewal by uplift, and is thus cyclical rather than linear.

The cyclic concepts of Davis have been thoroughly debated and criticized—and applied (see, e.g., Chorley et al., 1973; Ritter, 1988; Thorn, 1988; Beckinsale & Chorley, 1991). Davisian geomorphology fell out of fashion in the mid twentieth century, but implicit invocations of peneplains and Davisian cycles persist in historical geology, and Davis' explanatory triad of structure, process, and time is firmly entrenched in Earth science. Under certain (uncommon) circumstances, the Davisian sequence is a logical outcome (Gournellos, 1997), and the interpretation of erosion surfaces (peneplains) remains an important component of historical geology and geomorphology. Lester C. King (1953, 1957) presented a cyclical model that incorporated some ideas of both Davis and Penck (see below). King's model, first applied to and inspired by landscapes of southern Africa, involves the parallel retreat of slopes (in contrast to the downwasting of Davis' model) ultimately producing a pediplain, and was linked to broad-scale crustal deformation.

The model of Walther Penck (1924) is widely misunderstood and misinterpreted, and was historically important in the Anglophone world mainly as an alternative to the cycle of erosion (Thorn, 1988). However, a key contribution was the recognition that uplift and denudation operate simultaneously, with relief a function of their relative importance rather than of stages in a cycle. Through his "renewal of exposure" concept, Penck (1924) also articulated a feedback included in many landscape evolution models—increased rock weathering rates as overlying material is eroded away.

An alternative to cyclic models was presented by John T. Hack (1960), who linked it to concepts presented by Grove Karl Gilbert (1877). This concept of dynamic equilibrium is a misnomer from a systems theory perspective, suggesting that in an erosionally graded landscape, every slope and stream channel is adjusted to every other, with all eroding at the same rate (Hack, 1960). The evolutionary assumption underlying "dynamic equilibrium" (better characterized as steady state) is that landscape evolution tends toward this steady-state condition, with topography adjusted to characteristic forms adapted to the climate and geologic framework.

Principles of gradient, resistance, and thermodynamic selection could all readily be applied to the denudation sequences postulated by Davis and King, the oppositional dynamics of Penck's framework, and Hack's progression toward steady state. However, no selection principles seem to apply directly to these theories as opposed to the processes operating within them, except that perhaps resistance selection for stability could be applied to the steady state (dynamic equilibrium) goal function.

9.2.3 Equilibrium

Equilibrium is used here to encompass various ideas that the attractor states of landscape evolution represent characteristic forms or states that are adjusted to their environmental context, dynamically stable, and self-maintaining to some extent. This is distinct from short term chemical or thermal equilibria that may be achieved in mass and energy fluxes and transformations.

To a certain degree, equilibrium ideas can be linked to human psychology and history; related to desires for or notions of a balance or harmony of nature, and a sense of order. However, equilibria in ESS are not just a figment of our imaginations. They do exist, and in many cases equilibrium assumptions about attractors fit well with both theory and empirical knowledge. Further, some equilibrium notions are useful independently of their applicability to real landscapes, as benchmarks or simplifying modeling assumptions.

In ESS there often exist relaxation time equilibria. Responses to disturbances or change, or developmental pathways more generally, are finite and not infrequently decelerate, so that changes are more rapid early on, slow down rapidly, and eventually halt or decline to negligible rates. The resulting relaxed states are often labelled as equilibrium, though they are better seen as *apparent* attractors.

Another type of apparent attractor relates to stream profiles. A concave-up longitudinal profile (a plot of channel elevation vs. distance from the drainage divide or another upstream location to the stream mouth) has long been claimed or assumed to be an equilibrium form. This form has been supposed to represent the adjustment of channel slopes to transport the available sediment load. Stream power at a cross-section is

$$\Omega = \rho g Q S, \tag{9.6}$$

where the variables on the right are, respectively, water density, the gravity constant, discharge, and slope. In upstream reaches flow (Q) is lower, so a steeper slope is

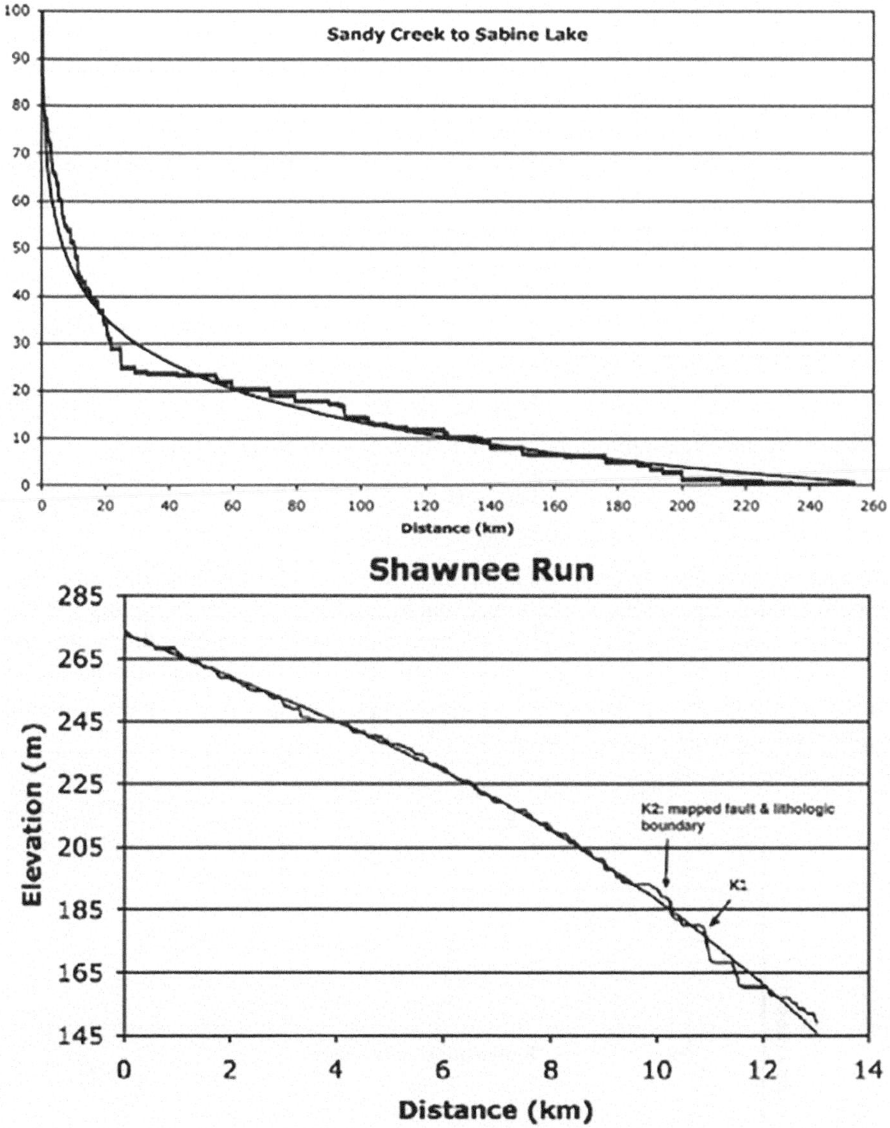

Fig. 9.2 Longitudinal profiles. Top is a classic concave-up profile from Louisiana; bottom is a convex profile from central Kentucky. In both profiles the actual digital elevation model-based profile is the irregular line, and the smooth line is a fitted profile. K1, K2 are knickpoints

necessary to move sediment. Further downstream, as Q increases, a lower slope is required producing the concave slope (Fig. 9.2). There are numerous difficulties with this concept, as outlined elsewhere (e.g., Goldrick & Bishop, 2007; Harmar & Clifford, 2007; Phillips & Lutz, 2008; Khan et al., 2021).

The constrained gradients concept explains the general tendency toward concave profiles, and the numerous deviations therefrom, as an emergent outcome of gradient selection and force-resistance relationships. Elevation at the drainage divide and at the stream mouth is given by H_A, H_B. The longitudinal profile is dH/dx, where x is flow distance from A to B. Though local gradients may be vertical, overhanging, or negative (e.g., at the downstream end of pools), in general channel slope is constrained by the maximum steepness the channel material can maintain (S_{max}) and a minimum gradient required to move water (S_{min}). For any segment of the channel profile,

$$S_{\min} \leq dH/dx \leq S_{\max} \tag{9.7}$$

L is the A-B stream length, and in most cases the minimum slope is insufficient to get water to the base level at H_B:

$$H_A - S_{\min} L > H_B \tag{9.8}$$

Similarly, the maximum slope will not allow flow to cover the distance:

$$\frac{(H_A - H_B)}{S_{\max}} < L) \tag{9.9}$$

The maximum convexity profile is a hanging valley type profile, with the minimum slope from drainage divide to the lower reaches, where it descends rapidly via the maximum slope to base level at H_B. The maximum concavity profile drops at slope S_{max} from the drainage divide to an elevation allowing the minimum slope to be maintained to the outlet. These profiles define a possibility space within which any segments that satisfy eq. (9.7) can occur (Fig. 9.3).

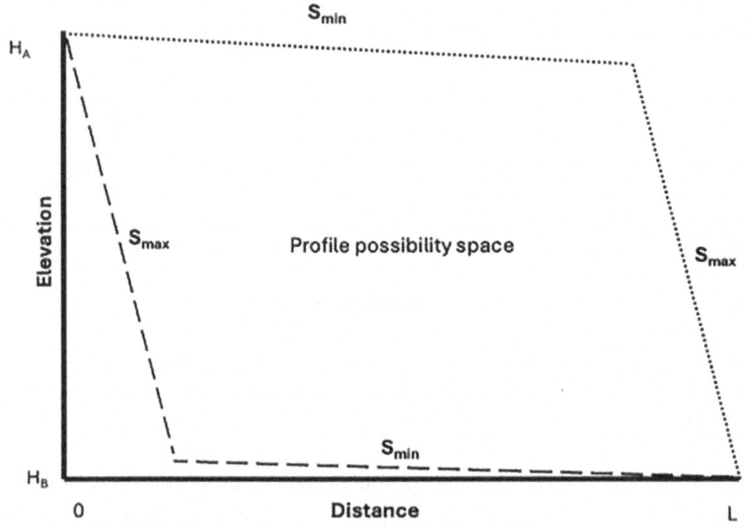

Fig. 9.3 The profiles of maximum concavity (lower, dashed line) and convexity (upper, dotted line) between the drainage divide and stream mouth, assuming a maximum slope constrained by geomechanical stability (S_{max}) and a minimum channel slope (S_{min}) needed to convey flow. Actual profiles fall into the possibility space between the extremes

Erosion along the channel length pushes the profile toward the maximum concavity limit, as does upstream migration of knickpoints. Further, gradient selection directs flows along steeper local gradients. Thus, increasing concavity simply reflects channel erosion where shear stress is sufficient, with the many deviations from concavity often attributable to resistance and gradient selection across variable materials and structures.

The evolution of profiles towards concavities—when and where it occurs—is better and more simply explained as an emergent outcome of gradient and resistance selection within slope constraints than as an effort by streams to balance sediment supply and transport capacity.

9.2.4 Power-law Distributions

Power laws—we have encountered many already in this book—are relationships of the form $y = aX^b$. Power law distributions are a subclass of heavy-tailed distributions. The latter are probability distributions with tails that are not exponentially bounded, meaning that large events (extreme values of the distribution) are more likely compared to distributions with lighter tails, such as the normal distribution. Heavy-tailed distributions are ubiquitous in nature—a partial list from Guadagnini and Neuman (2011) includes: atmospheric turbulence, diffusion-limited aggregation, kinetic surface roughening, fluvial turbulence, sand wave dynamics, Martian topography, river morphometry, gravel-bed mobility, barometric pressure, low-energy cosmic rays, cosmic microwave background radiation, metal-insulator transitions, irregularities in human heartbeat time series, and turbulent boundary layers of the Earth's magnetosphere. Other well-known examples include fire, landslide, earthquake, and city size distributions (Fig. 9.4). They are so common in the geosciences that they can be considered a potential attractor state.

Power laws are the most common heavy-tailed distribution sought, found, and assumed in the geosciences, due to their ease of interpretation compared to other heavy-tailed distributions, their connections to other theoretical and analytical frameworks (e.g., fractals, chaos, self-organized criticality, $1/f$ noise, Hurst phenomena), and, in many cases, their consistency with geophysical logic. To the extent there is debate over power-law distributions in the geosciences it is mainly over methods for estimating the coefficient a and exponent b, whether other heavy-tailed distributions provide a better fit to data, and their implications. For example, power law distributions of flood magnitudes imply more frequent large floods than would otherwise be expected.

Back in 2015, in a blog I no longer maintain, I published a post called "The Dubious Power of Power Laws" (archived here: https://geography.as.uky.edu/blogs/jdp?page=15, last accessed 19 Sept. 2024). In it I argued that power laws are an example of equifinality (where different processes or causes can produce the same or very similar outcomes). Therefore, identifying a power law distribution, *by itself*, has no meaning other than that a power law can describe the distribution—you could say the same about any statistical finding. But some interpreted my stance as

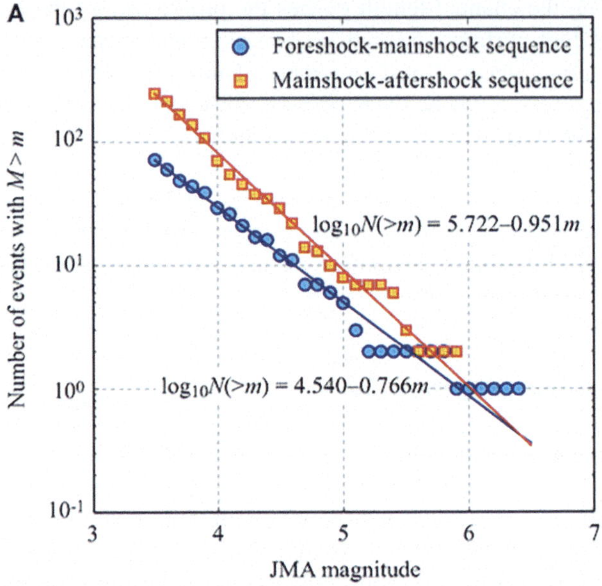

Fig. 9.4 Characteristics of foreshock-mainshock and mainshock-aftershock sequences of the 2016 Kunamoto earthquakes according to Gutenberg–Richter power law model. (Reprinted under CC-BY-4.0 from Goda et al. (2016). © The Authors)

being that power laws are a purely statistical artifact and are not a real feature of nature. That was not my intent, and clearly numerous natural processes can produce power law (and other heavy-tailed) distributions. The key point is the "numerous natural processes"—without some additional evidence, a power-law distribution does not tell us which of the numerous processes is (or are) involved.

Mathematically, heavy-tailed, right-skewed distributions are difficult to tell apart. One study reanalyzed examples of purported power law distributions and found that the evidence that a power function is the best fit is inconclusive in many cases (Clauset et al., 2009). Mitzenmacher (2004) showed that lognormal distributions in particular are difficult to distinguish from power laws in empirical data. Other studies also show that apparent power laws may disappear when subjected to more stringent testing (e.g., Lima-Mendez & van Helden, 2009). $1/f$ (f is for frequency) noise is a type of power law distribution often found in hydrology and climatology, and Li and Zhao (2012) outlined three different theorems for producing it. Martin and Monnier (2015) explored some specific equifinality problems in inferring properties of pseudoplastic geophysical flows from power laws. Of 11 hypotheses to explain heavy tails of flood peak distributions they evaluated, Merz et al. (2022) found 6 to have high plausibility.

More importantly, though, is that even when power law distributions are real, they don't (again, by themselves) necessarily tell us anything other than that the distribution follows a power law. The need to go beyond fitting power law distributions for scientific explanation is emphasized by Pinto et al. (2012). This involves linking the distributions to specific causal mechanisms, and more detailed distinctions among such distributions. For example, Corral and González (2019) reanalyzed examples of proposed power law distributions dealing with potentially damaging natural phenomena, where adequate fits of the distributions of sizes are especially important in assessing hazards. Employing a new statistical fitting method, they clarified the range of validity of the power law fit and the corresponding exponent and whether a power law tail is improved by a truncated lognormal. They found that impact fireballs and California earthquakes show untruncated power law behavior, but that global earthquakes follow a double power law. Rainfall and tropical cyclones show a truncated power law regime. Truncated lognormals, by constrast, better described karst sinkhole and wildfire size distributions. Carroll (1982) identified five different classes of models (each with multiple specific models) that can potentially explain city rank-size distribution power laws. He also found that many of these are plausible, but directly contradict each other. Mitzenmacher (2004), from a computer science perspective, identified five classes of models that generate power law or closely related lognormal distributions.

So, what kind of phenomena can produce or explain power laws? Cascades of large-scale fluxes to successively smaller scales can account for 1/f noise (Gupta & Waymire, 1989; Newman & Turcotte, 1990), and storage effects in hydrological and geomorphic systems are another highly plausible and physically realistic explanation for Hurst effects and 1/f noise in landscapes (Kirkby, 1987). On geological time scales, Earth evolution is characterized by relatively long periods of stasis interspersed with relatively short periods of intense activity and change. Modeling this kind of tempo as a Cantor dust can produce both the fractal topography exhibiting 1/f noise often found in digital terrain models and the temporal sequences of uplift and denudation events in the geological record (Snow, 1992).

Another possible source of power law behavior is *scale invariance* and self-similarity. If similar form-process relationships occur across a range of spatial scales, resulting in self-similarity and fractal geometry, this will produce power-law distributions. *Deterministic chaos* is also associated with fractal geometry and power-law distributions. So is *self-organized criticality* (SOC), where systems (both real and "toy" hill slopes are a commonly used example) evolve toward critical thresholds. Given the threshold dominance of many geoscience phenomena, the attractiveness of this perspective in our field is obvious. Power law distributions have often been used as proof, or supporting evidence, for chaos, fractals, and SOC phenomena, but power law distributions are a necessary, but by no means sufficient, indicator.

Power laws also arise from *preferential attachment* phenomena. In economics this is manifested as the rich get richer; in internet studies as the tendency of highly linked sites to get ever more links and hits. Preferential attachment models have been applied in geography, evolutionary biology, geomicrobiology, soil science, hydrology, and geomorphology. Various *optimization* schemes, based on minimizing costs, entropy, etc. can also produce power laws. These have been linked to power laws quite extensively in studies of channel networks and drainage basins, as well as other geophysical phenomena.

Multiplicative cascade models—fractal or multifractal patterns arising from iterative random processes—produce power laws. Multiplicative cascades are widely applicable to phenomena such as rainfall, runoff, and river flows and have also been applied in fluid dynamics, soil physics, and geochemistry. And speaking of randomness, Mitzenmacher (2004) showed how monkeys typing randomly could produce power law distributions of word frequencies.

Diffusion limited aggregation is a process whereby particles (or analogous objects) undergoing random motion cluster together to form aggregates. The size distribution of the aggregates follows a power law. DLA has been used to model evolution of drainage networks, escarpments, and eroded plateaus, and applied in several other areas of geosciences.

Rather than the ubiquity of power law patterns representing some universal underlying law of nature, they seem to represent an emergent pattern that can arise from several different causes—*equifinality*. Thus, their meaning must be assessed case by case.

For example, estimated sediment deposition rates decrease as a power law function of the measurement interval. The apparent decrease has been attributed to the nature of the sedimentary record: as averaging time is lengthened, longer depositional hiatuses are included. Schumer and Jerolmack (2009) showed that a heavy-tailed distribution of hiatuses produces this effect, and that if geometric and geologic constraints provide upper-limit constraints on how long a hiatus can last, then average accumulation rates approach a constant value over very long times.

Selection can produce, or be associated with, power law distributions. Selection for minimum-resistance flow paths, for example, can be described by the power-law relations from percolation theory as described in Chaps. 6 and 8. But as with some of the other constructs discussed above, selection applies to the mechanisms within the system rather than the overall power law structure.

9.3 Multiple Path, Multiple Outcome Concepts

9.3.1 Alternative Stable States

A seesaw metaphor conveys the gist of alternative stable state concepts. The seesaw has two potential stable states (one side or other on the ground), separated by multiple possible in-between states that are unstable and transient. Like a seesaw, a key threshold or tipping point often separates the stable states. This could apply to the global climate, where there is some evidence of flipping between "icehouse" or even "snowball" states to "hothouse" or "greenhouse" states. However, there is evidence that there are multiple icehouse and hothouse states, and several factors that may trigger transitions (Summerhayes, 2020).

Alluvial rivers may have alternative states of net aggradation (deposition) or degradation (erosion and incision). The threshold or tipping point is where sediment supply is equal to sediment transport capacity (on a long-term average rather than instantaneous or event basis). Lane (1955) expressed this using an analytical balance as a visual metaphor. On one side is flow or discharge Q and slope S, representing transport capacity; on the other is sediment discharge Q_s and a measure of sediment size or caliber (D_{50}, the median particle diameter). Lane's relation is given as

$$Q_s D_{50} \propto Q S, \tag{9.10}$$

The ∝ sign is necessary because additional parameters are required to produce equivalent units. The Lane relationship shows that stream degradation and aggradation respond to changes in the relationship between sediment supply and transport capacity. This has often been interpreted within steady-state equilibrium paradigms to suggest that streams somehow manipulate Q_s, Q_w, D_{50}, and (especially) S to keep the balance precisely in the middle. A more realistic interpretation is that if the system *is* precisely balanced (in steady state), a change in any factor will tip the scales one way or another. Further, if transport capacity is much greater or less than sediment supply, it would require a great deal of change to achieve steady state or tip the scale or seesaw back the other way. Most alluvial channels are indeed observed to experience net aggradation or degradation, while steady state is relatively rare, except in a very approximate sense.

Alternative stable states are in most cases, like the example above, associated with threshold-dominated systems. Indeed, one could identify alternative states for runoff production, glacier advances or retreats, state changes of water, and innumerable other phenomena. The primary utility of the concept is to present alternatives to single-path, single-outcome frameworks. As such, it has been particularly utilized in ecology.

9.4 Goal functions and Optimality

Many goal functions have been proposed for climate, hydrological, geomorphic, and ecosystems based on thermodynamic principles and criteria. Most of these are mutually consistent, and equivalent in terms of mathematical formulations and thermodynamic implications. Table 9.2 lists some of the optimality-based goal functions for abiotic processes in ESS; additional examples in ecological and biological systems are given by Patten (1995) and Fath et al. (2001).

9.4.1 Goal functions and Emergence

Scientific explanations or hypotheses should not require intentionality of non-sentient entities or an external controller (though they need not necessarily exclude or forbid one either).

Rather than a universal physical law or geographical principle dictating optimality, sufficient conditions for optimality to emerge are that higher probabilities of optimal behavior are associated with positive feedbacks. Optimal-like behavior in landscape systems does not require any goal functions. All that is necessary is that

Table 9.2 Optimal principles and goal functions based on, or implying, maximum entropy production (MEP) or the least action principle (LAP). All are examples of thermodynamic selection

Principle	Domain	Summary	Comments	Source
Maximum generation of available potential energy	Atmosphere	Atmosphere heat flux operates to maximize rate of potential energy production	Equivalent to maximum entropy production	Lorenz 1960
Minimum stream power	Fluvial channels	Channels adjust to transport sediment with minimum possible expenditure of energy	Phrased in various forms	e.g., Brebner & Wilson, 1967; Yang, 1971; reviews: Griffiths, 1984; Paik & Kumar, 2010
Maximum flow efficiency	Fluvial channels	Channels adjust to maximize flow efficiency & minimize energy expenditure	Phrased in various forms	e.g., Davies & Sutherland, 1980; Yang et al., 1981; Jia, 1990 reviews: Molnar & Ramirez, 1998; Paik & Kumar, 2010
Minimum entropy exchange	Climate; atmospheric circulation	Ocean-atmosphere heat flux minimizes entropy exchange with external environment	Equivalent to maximum entropy export to external environment	Paltridge, 1975
LAP	Fluvial channels	Channels tend to adjust to transport water & sediment with the minimum possible available energy	Consistent with maximum flow efficiency	Huang & Nanson, 2000; Nanson & Huang, 2008, 2017
MEP	Turbulent flows	At steady state, entropy export maximized by turbulent energy dissipation		Ozawa et al., 2003; Dewar, 2005
MEP	Environmental systems	Nonequilibrium thermodynamic systems organized in steady state such that entropy production is maximized	Applied to multiple geophysical, ecological, & hydrological systems	Kleidon et al., 2010
Maximum power	Drainage basin evolution	Maximization of sediment transport to deplete topographic gradients	Based on tendency of ESS to deplete driving gradients as rapidly as possible	Kleidon et al., 2013

(continued)

Table 9.2 (continued)

Principle	Domain	Summary	Comments	Source
LAP	Geomorphology	Geomorphic work accomplished using minimal amount of available energy	In fluvial channels equivalent to maximum flow efficiency	Nanson & Huang, 2018
MEP	Atmospheric circulation	General global circulation & heat fluxes explained by energy gradient depletion guided by Carnot limit	Discussed in Chap. 7	Kleidon, 2023

Fig. 9.5. A barchan dune at Great Sand Dunes National Park and Preserve (Colorado, USA). (U.S. National Park Service photo by Patrick Myers)

the maximization or minimization involved increases the odds of survival and replication of the system involved. Optimality in ESS is not teleological or deterministically inevitable. Rather, it is an emergent property arising from selection.

Consider the example of the crescent-shaped aeolian sand dunes with the horns pointed downwind known as barchans (Fig. 9.5). Barchans are found in sandy arid regions with a wind regime characterized by a single dominant direction, abundant but not unlimited sand supply, and little or no vegetation cover.

Barchans were traditionally interpreted as steady state features with input and losses of sand approximately balanced, and roughly constant size and shape even as the dune migrates downwind. However, studies in recent decades indicate that size-shape invariance is not necessarily common (e.g., Sauermann et al., 2000; Kroy et al., 2005; El belhriti & Douady, 2011). A degree of geometric consistency does exist, however, and laboratory studies suggest that a classic barchan state is an attractor (Groh et al., 2009).

Werner (1995) devised a cellular automata model of aeolian dunes, based on three simple rules: sand saltation occurs only in the direction of the wind, sand deposition is more probable where mobile sand already exists than on bare ground, and an angle of repose exists on the downwind face. These rules do not preordain any specific dune morphology, but under the wind direction, sand supply, and vegetation-free conditions described above, barchan forms develop. This indicates that the attractor form is an emergent property, a finding verified by later work of Kocurek and Ewing (2005).

Gradient selection and threshold-mediated modulation explain the emergent pseudo-equilibrium form of barchans (Phillips, 2011). Gradient selection links the downwind-oriented crescentic shape to the interactions among transport rates, crest height, and sand availability (Hesp & Hastings, 1998; Groh et al., 2009; El belrhiti & Douady, 2011). Slipface gradients are limited by the angle of repose, and sand avalanching down this slope gradient limits crest heights for a given sand mass. Gradient selection and topographic steering also drive feedbacks among crest elevation, divergence of streamlines near the upwind toe, shear stress at the crest, and sideslope sand delivery and transport (Hesp & Hastings, 1998). This results in an increase in shear stress at the crest, reducing height by sand deflation. For a given dune width, these thresholds related to gradient selection maintain heights within a narrow range.

The emergence of barchans is sometimes interpreted or described as the dune attempting to seek and maintain a steady-state equilibrium (including some of the references cited above). But because the characteristic forms emerge without any goal functions, the emergent explanation is simpler and satisfies Ockham's Razor (discussed further below).

9.4.2 Circles, Spheres, and Hemispheres

Here we examine a recurrent geometry in ESS from the perspective of attractors and selection. The circle, and its three-dimension relatives, the sphere and hemisphere, are common in landscapes—impact craters from meteors to raindrops, explosion craters, sinkholes, weathering cavities, bubbles, center-pivot irrigation areas, and tree canopy footprints in the form of driplines, to name a few.

A sphere is the most efficient shape for enclosing a given volume in terms of surface-to-volume ratio, and a circle the most economical way to enclose an area, with the smallest perimeter/area ratio. These are thus selected for. For example, surface tension acting to pull molecules into the tightest possible grouping forms spheres, and the effects of these spheres (water drops, bubbles) tends to be approximately circular or hemispherical (e.g., cavitation pits in rock). Irrigating the maximum area with the minimum amount of pipe points to circular center-pivot irrigation when the terrain allows it.

Multidirectional isotropic growth or dispersal from a point also produces circular patterns. When neither sunlight nor space for growth is obstructed, tree branches grow away from the trunk with an equal probability in any direction (same for roots below the ground for trees with lateral roots). A circular pattern as projected straight up (or defined by the outermost branch tips) covers the most area with the least effort for capturing sunlight or water. Capturing the maximum insolation with

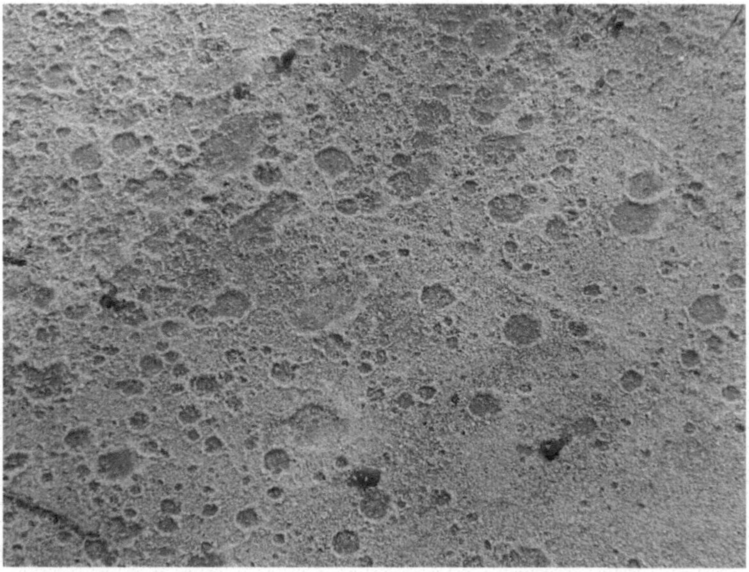

Fig. 9.6 Raindrop impact craters in sand

minimum growth effort should be favored by preferential selection. The area of influence on the ground reflected by driplines, litter fall, and soil moisture drawdown is roughly circular.

Point-centered disturbances without any directional bias also produce more or less circular perimeters—explosion craters from volcanic eruptions or bombs, impact craters, and sinkholes centered on solution or failure at a point (e.g., an intersection of vertical joints or a karst shafts) (Fig. 9.6). No advantage to formation of circular shapes for these cases has been proposed, though in some cases mechanical stability may promote their preservation via resistance selection, and ejection of material in a circular area is the most efficient way (in terms of mass/area) of displacing of the ejected mass.

It could be argued in all these cases that convergence to approximately spherical or circular shapes represents a general selection principle for geometric efficiency in, e.g., encompassing space, distributing displaced material, or acquiring isotropically distributed resources. This is perhaps a more obvious example of emergence from selection, as opposed to asserting that nature prefers circles and spheres.

9.5 Of Razors and Emergence

Selection is *emergent*; emergence is a simpler explanation than the alternatives, and it works. Therefore, it is preferred by Ockham's Razor, attributed to English philosopher and theologian William of Ockham (1287–1347): which is often rendered as: simpler explanations are more likely to be correct, or if two explanations work equally well with respect to explanation and consistency with available evidence, the simpler of the two is preferred.

Forms or patterns that recur repeatedly in nature suggest that they might be goals of nature or the expression of a law of nature. But we are not obliged to explain such features in those terms when they can be explained in simpler emergent terms, as Ockham's Razor tells us we should do. If similar phenomena can emerge from a few simple rules or constraints, and if these are applicable in multiple locations, then we should expect them to emerge repeatedly.

References

Beckinsale, R. P., & Chorley, R. J. (1991). The history of the study of landforms or the development of geomorphology. In *Historical and regional geomorphology 1890–1950* (Vol. 3). Routledge. 496 pp.

Bestelmeyer, B. T., Tugel, A. J., Peacock, D. G., et al. (2009). State-and-transition models for heterogeneous landscapes: A strategy for development and application. *Range Ecology and Management, 62*, 1–15.

Brebner, A., & Wilson, K. C. (1967). Derivation of the regime equations from relationships for pressurized flow by use of the principle of minimum energy-degradation rate. *Proceedings, Institute of Civil Engineers, 36*, 47–62.

Briske, D. D., Fulendor, S. D., & Smeins, F. E. (2005). State-and-transition models, thresholds, and rangeland health: A synthesis of ecological concepts and perspectives. *Range Ecology and Management, 58*, 1–10.

Carroll, G. R. (1982). National city-size distribution—What do we know after 67 years of research? *Progress in Human Geography, 6*, 1–43.

Cattelino, P. J., Noble, I. R., Slayter, R. O., & Kessell, S. R. (1979). Predicting the multiple pathways of plant succession. *Environmental Management, 3*, 41–50.

Chorley, R. J., Beckinsale, R. P., & Dunn, A. J. (1973). *The history of the study of landforms or the development of geomorphology: Vol. 2. The Life and Work of William Morris Davis* (p. 874). Methuen.

Clauset, A., Shalizi, C. R., & Newman, M. E. J. (2009). Power-law distributions in empirical data. *SIAM Review, 51*, 661–671.

Clements, F. E. (1916). *Plant succession. An analysis of the development of vegetation*. Carnegie Institution, Pub. No. 242.

Corral, A., & González, A. (2019). Power law size distributions in geoscience revisited. *Earth and Space Science, 6*, 673–697.

Cowles, H. C. (1899). The ecological relations of the vegetation on the sand dunes of Lake Michigan. *Botanical Gazette, 27*, 95–117, 167–202, 281–308, 361–391.

Davies, T. R. H., & Sutherland, A. J. (1980). Resistance to flow past deformable boundaries. *Earth Surface Processes and Landforms, 5*, 175–179.

Davis, W. M. (1899). The peneplain. *American Geologist, 23*, 207–239.

Davis, W. M. (1909). *Geographical essays*. Ginn.

Dewar, R. C. (2005). Maximum entropy production and non-equilibrium statistical mechanics. In A. Kleidon & R. D. Lorenz (Eds.), *Non-equilibrium thermodynamics and the production of entropy. Understanding complex systems* (pp. 41–55). Springer.

Dokuchaev, V. V. (1883). Russian Chernozem. Selected works of V.V. Dokuchaev (Vol. 1, pp. 14–419). Moscow, 1948. Israel Program for Scientific Translations Ltd. (for USDA-NSF), S. Monson, Jerusalem, 1967. (Translated from Russian by N. Kaner).

El Belrhiti, H., & Douady, S. (2011). Equilibrium versus disequilibrium of barchan dunes. *Geomorphology, 125*, 558–568.

Fath, B. D., Patten, B. C., & Choi, J. S. (2001). Complementarity of ecological goal functions. *Journal of Theoretical Biology, 208*, 493–506.

Gersmehl, P. J. (1976). An alternative biogeography. *Annals of the Association of American Geographers, 66*, 223–241.

Gilbert, G. K. (1877). *Report on the geology of the Henry Mountains*. U.S. Geographical and Geological Survey.

Gleason, H. A. (1939). The individualistic concept of the plant association. *American Midland Naturalist, 21*, 92–110.

Goda, K., Campbell, G., Hulme, L., et al. (2016). The 2016 Kumamoto earthquakes: Cascading geological hazards and compounding risks. *Frontiers in Built Environment, 2*, 19. https://doi.org/10.3389/fbuil.2016.00019

Goldrick, G., & Bishop, P. (2007). Regional analysis of bedrock stream long profiles: evaluation of Hack's SL form, and formulation of an alternative (the DS form). *Earth Surface Processes and Landforms, 32*, 649–671.

Gournellos, T. (1997). A theoretical Markov chain model of the long term land form evolution. *Zeitschrift für Geomorphologie, 41*, 519–529.

Griffiths, G. A. (1984). Extremal hypotheses for river regime: An illusion of progress. *Water Resources Research, 20*, 113–118.

Groh, C., Rehberg, I., & Kruell, C. A. (2009). How attractive is a barchan dune? *New Journal of Physics, 11*. https://doi.org/10.1088/1367-2630/11/2/023014

Guadagnini, A., & Neuman, S. P. (2011). Extended power-law scaling of self-affine signals exhibiting apparent multifractality. *Geophysical Research Letters, 38*, L13403.

Gupta, V. K., & Waymire, E. (1989). Multiscaling properties of spatial rainfall and river flow distributions. *Journal of Geophysical Research, 95D*, 1999–2009.

Hack, J. T. (1960). Interpretation of erosional topography in humid temperate regions. *American Journal of Science, 285*, 80–97.

Harmar, O. P., & Clifford, N. J. (2007). Geomorphological explanation of the long profile of the lower Mississippi River. *Geomorphology, 84*, 222–240.

Hesp, P. A., & Hastings, K. (1998). Width, height, and slope relationships and aerodynamic maintenance of barchans. *Geomorphology, 22*, 193–204.

Huang, H. Q., & Nanson, G. C. (2000). Hydraulic geometry and maximum flow efficiency as products of the principle of least action. *Earth Surface Processes and Landforms, 25*, 1–17.

Jenny, H. (1941). The Factors of Soil Formation. *A System of Quantitative Pedology*. Dover.

Jia, Y. (1990). Minimum Froude number and the equilibrium of sand bed rivers. *Earth Surface Processes and Landforms, 15*, 199–209.

Johnson, D. L., & Watson-Stegner, D. (1987). Evolution model of pedogenesis. *Soil Science, 143*, 349–366.

Khan, S., Fryirs, K. A., & Ralph, T. J. (2021). Geomorphic controls on the diversity and patterns of fluvial forms along longitudinal profiles. *Catena, 203*, 105329.

King, L. C. (1953). Canons of landscape evolution. *Geological Society of America Bulletin, 64*, 721–751.

King, L. C. (1957). The uniformitarian nature of hillslopes. *Transactions of the Edinburgh Geological Society, 17*, 81–102.

Kirkby, M. J. (1987). The Hurst effect and its implications for extrapolating process rates. *Earth Surface Processes and Landforms, 12*, 57–67.

Kleidon, A. (2023). Working at the limit: a review of thermodynamics and optimality of the Earth system. *Earth System Dynamics 14*, 861–896.

Kleidon, A., Malhi, Y., & Cox, P. M. (2010). Maximum entropy production in environmental and ecological systems. *Philosophical Transactions of the Royal Society B, 365*, 1297–1302.

Kleidon, A., Zehe, E., Ehret, U., & Scherer, U. (2013). Thermodynamics, maximum power, and the dynamics of preferential river flow structures at the continental scale. *Hydrology and Earth System Sciences, 17*, 225–251.

Kocurek, G., & Ewing, R. C. (2005). Aeolian dune field self-organization—Implications for the formation of simple versus complex dune-field patterns. *Geomorphology, 72*, 94–105.

Kroy, K., Fischer, S., & Obermayer, B. (2005). The shape of barchan dunes. *Journal of Physics: Condensed Matter, 17*, S1229–S1235. https://doi.org/10.1088/0953-8984/17/14/012

Lane, E. W. (1955). The importance of fluvial morphology in hydraulic engineering. *Proceedings of the American Society of Civil Engineers, 81*, 1–17.

Li, L., & Zhao, W. (2012). On 1/f noise. *Mathematical Problems in Engineering, 2012*, 673648.

Lima-Mendez, G., & van Helden, J. (2009). The powerful law of the power law and other myths in network biology. *Molecular BioSystems, 5*, 1482–1493.

Lorenz, E. N. (1960). Generation of available potential energy and the intensity of the general circulation. In R. F. Pfeffer (Ed.), *Dynamics of climate* (pp. 86–92). Permagon.

Lorenz, E. N. (1963). Deterministic nonperiodic flow. *Journal of Atmospheric Science, 20*, 130–141.

Marbut, C. F. (1935). *Soils of the United States, Part III, Atlas of American agriculture*. U.S. Department of Agriculture.

Martin, N., & Monnier, J. (2015). Inverse rheometry and basal properties influence for pseudoplastic geophysical flows. *European Journal of Mechanics B-Fluids, 50*, 110–126.

Merz, B., Basso, S., Fischer, S., et al. (2022). Understanding heavy tails of flood peak distributions. *Water Resources Research, 58*, e2021WR030506.

Mitzenmacher, M. (2004). A brief history of generative models for power law and lognormal distributions. *Internet Math., 1*, 226–251.

Molnar, P., & Ramirez, J. A. (1998). Energy dissipation theories and optimal characteristics of river networks. *Water Resources Research, 34*, 1809–1818.

Mori, A. S. (2011). Ecosystem management based on natural disturbances: hierarchical context and non-equilibrium paradigm. *Journal of Applied Ecology, 48*, 280–292.

Nanson, G. C., & Huang, H. Q. (2008). Least action principle, equilibrium states, iterative adjustment and the stability of alluvial channels. *Earth Surface Processes and Landforms, 33*, 923–942.

Nanson, G. C., Huang, H. Q. (2017). Self-adjustment in rivers: Evidence for least action as the primary control of alluvial-channel form and process. *Earth Surface Processes and Landforms 42*, 575–594.

Nanson, G. C., & Huang, H. Q. (2018). A philosophy of rivers: equilibrium states, channel evolution, teleomatic change and the least action principle. *Geomorphology, 302*, 3–19.

Newman, W. I., & Turcotte, D. L. (1990). Cascade model for fluvial geomorphology. *Geophysical Journal International, 100*, 433–439.

Nikiforoff, C. C. (1949). Weathering and soil evolution. *Soil Science, 67*, 219–230.

Ozawa, H., Ohmura, A., Lorenz, R. D., & Pujol, T. (2003). The second law of thermodynamics and the global climate system: a review of the maximum entropy production principle. *Reviews of Geophysics, 41*, 1018. https://doi.org/10.1029/2002RG000113

Paik, K., & Kumar, P. (2010). Optimality approaches to describe characteristic fluvial patterns on landscapes. *Philosophical Transactions of the Royal Society B, 365*, 1387–1395.

Paltridge, G. W. (1975). Global dynamics and climate—A system of minimum entropy exchange. *Quarterly Journal of the Royal Meteorological Society, 101*, 475–484.

Patten, B. C. (1995). Network integration of ecological extremal principles: exergy, emergy, power, ascendency, and indirect effects. *Ecological Modelling, 79*, 75–84.

Penck, W. (1924). *Morphological analysis of landforms*. English translation (Czech H, Boswell KC, translators) published 1972. Hafner.

Pinto, C. M. A., Mendes Lopes, A., & Tenreiro Machado, J. A. (2012). A review of power laws in real life phenomena. *Communications in Nonlinear Science and Numerical Simulation, 17*, 3558–3578.

Phillips, J. D. (1993a). Instability and chaos in hillslope evolution. *American Journal of Science, 293*, 25–48.

Phillips, J. D. (1993b). Progressive and regressive pedogenesis and complex soil evolution. *Quaternary Research, 40*, 169–176.

Phillips, J. D. (1993c). Stability implications of the state factor model of soils as a nonlinear dynamical system. *Geoderma, 58*, 1–15.

Phillips, J. D. (1998). On the relations between complex systems and the factorial model of soil formation (with discussion and response). *Geoderma, 86*, 1–43.
Phillips, J. D. (1999). *Earth surface systems. Complexity, order, and scale*. Basil Blackwell.
Phillips, J. D. (2006). Deterministic chaos and historical geomorphology: A review and look forward. *Geomorphology, 76*, 109–121.
Phillips, J. D. (2011). Emergence and pseudo-equilibrium in geomorphology. *Geomorphology, 132*, 319–326.
Phillips, J. D. (2017). Soil complexity and pedogenesis. *Soil Science, 182*, 117–127.
Phillips, J. D., & Lutz, J. D. (2008). Profile convexities in bedrock and alluvial streams. *Geomorphology, 102*, 554–566.
Phillips, J. D., & Van Dyke, C. (2017). Geomorphological state-and-transition models. *Catena, 153*, 168–181.
Pulsford, S. A., Lindenmayer, D. B., & Driscoll, D. A. (2014). A succession of theories: purging redundancy from disturbance theory. *Biological Reviews*. https://doi.org/10.1111/brv.12163
Ritter, D. F. (1988). Landscape analysis and the search for geomorphic unity. *Geological Society of America Bulletin, 100*, 160–171.
Sauermann, G., Rognon, P., Poliakov, A., & Herrmann, H. J. (2000). The shape of the barchan dunes of southern Morocco. *Geomorphology, 36*, 47–62.
Schumer, R., & Jerolmack, D. J. (2009). Real and apparent changes in sediment deposition rates through time. *Journal of Geophysical Research-Earth Surface, 114*, F00A06.
Simonson, R. W. (1959). Outline of a generalized theory of soil genesis. *Soil Science Society of America Proceedings, 23*, 152–156.
Simonson, R. W. (1978). A multiple-process model of soil genesis. In W. C. Mahaney (Ed.), *Quatenary soils* (pp. 1–25). GeoAbstracts.
Sivakumar, B. (2017). *Chaos in hydrology*. Springer.
Snow, R. S. (1992). The Cantor dust model for discontinuity in geomorphic process rates. *Geomorphology, 5*, 185–194.
Summerhayes, C. P. (2020). *Palaeoclimatology: From snowball earth to the anthropocene*. Wiley-Blackwell.
Tansley, A. G. (1935). The use and abuse of vegetational terms and concepts. *Ecology, 16*, 284–307.
Thorn, C. E. (1988). *Introduction to theoretical geomorphology*. Unwin-Hyman.
Thorpe, A. S., Achenhoug, E. T., Atwater, D. Z., & Callaway, R. M. (2011). Interactions among plants and evolution. *Journal of Ecology, 99*, 729–740.
Turcotte, D. L. (1997). *Fractals and Chaos in geology and geophysics* (2nd ed.). Cambridge University Press.
van der Walk, A. G. (2014). From formation to ecosystem: Tansley's response to Clements' climax. *Journal of the History of Biology, 47*, 293–321.
Vellend, M. (2016). *The theory of ecological communities*. Princeton University Press. 229 p.
Werner, B. T. (1995). Eolian dunes: Computer simulation and attractor interpretation. *Geology, 23*, 1107–1110.
Westoby, M., Walker, B., & Noy-Meir, I. (1989). Opportunistic management for rangelands not at equilibrium. *Journal of Range Management, 42*, 266–274.
Yang, C. T. (1971). Potential energy and stream morphology. *Water Resources Research, 7*, 311–322.
Yang, C. T., Song, C. C. S., & Woldenberg, M. J. (1981). Hydraulic geometry and minimum rate of energy dissipation. *Water Resources Research, 17*, 1014–1018.

Integrated Approaches to Earth Systems 10

Abstract

Multiple worldviews and scientific cultures apply to geosciences and are reflected in scientific practice. These are not mutually exclusive, and no logical or practical necessity prevents movement across and between them. Newtonian, Langrangian, Darwinian, Boltzmann, and von Bertalanffian approaches have all resulted in valid, useful advances in geophysics and Earth system science. There is no reason to reject or devalue any of these, and many good reasons to integrate them. The dominant controls and law-place-history frameworks are reflections of, and tools for, integration of scientific cultures, and both are derived directly from Earth sciences. ESS are characterized by multiple scale causality, so bottom-up, top-down, and scale-specific approaches are all applicable, along with explicit attention to issues of scale linkage and contingency. This also indicates a need to incorporate both reductionist and holistic perspectives. The chapter also shows that evolutionary creativity is not restricted to biota. Abiotic processes have led to emergence of novelties that are selected for because they enhance efficiency or survivability of ESS. Biological selection may continuously produce innovations in the form of new adaptations in different settings, but abiotic selection repeatedly reinvents similar adaptations. Climate science, writ large, and the case studies reviewed here (tropical cyclones, fluviokarst, and sandstone mesas) all illustrate abiotic creativity, and all rely on application of multiple, integrated worldviews.

J. D. Phillips, *Abiotic Selection in Earth Surface Systems*, Geophysics and Environmental Physics, https://doi.org/10.1007/978-3-031-85862-8_10

10.1 Worldviews

10.1.1 Darwin, Newton (and Laplace, Boltzmann, and von Bertalanffy)

In a widely cited essay in *Physics Today,* Harte (2002) contrasted the prevailing perspectives of physics vs. ecology: "*Physicists seek simplicity in universal laws. Ecologists revel in complex interdependencies. A sustainable future for our planet will probably require a look at life from both sides.*" Harte choose Isaac Newton and Charles Darwin as the symbolic figureheads for the physics and ecology worldviews. In physics, according to Harte, the more you look or the deeper you dig, the simpler it gets. Initial conditions are more important than historical factors, and the focus is on a search for universal patterns and laws. Physicists seek prediction and reserve a central role for ideal systems (e.g., ideal gas, noncompressible fluid). In ecology, by contrast, the deeper you dig and the more you look, the more complicated and complex it gets. History and contingency are critical, and law-seeking takes a back seat to case-by-case explanation. Ideal systems are seen, by Harte's ecologists, as caricatures of nature.

Readers, as well as those who cite Harte (2002) and Harte himself, recognize these as broad-brush generalizations with many exceptions, essentially as illustrative caricatures. But the desirability of synthesizing Darwinian and Newtonian worldviews is widely accepted as at least desirable, if not necessary, even by those who remain on one side or the other.

Beyond Newton and Darwin, other figureheads could be chosen to represent philosophies or worldviews relevant to and present in the study of Earth surface systems (ESS), such as Pierre-Simon Laplace (1749–1827), the French mathematician who made numerous contributions to physics and engineering as well as mathematics. Laplace can be nominated as a champion of deterministic causation:

> We may regard the present state of the universe as the effect of its past and the cause of its future. An intellect which at a certain moment would know all forces that set nature in motion, and all positions of all items of which nature is composed, if this intellect were also vast enough to submit these data to analysis, it would embrace in a single formula the movements of the greatest bodies of the universe and those of the tiniest atom; for such an intellect nothing would be uncertain and the future just like the past would be the present to it (Laplace 1814).

Thus, in the Laplacian scheme (like the others discussed here, this is a broad generalization or caricature), anything can be understood and explained if we just get enough information and know the relevant laws.

Austrian physicist Ludwig Boltzmann (1844–1906) represents a statistical mechanical view of nature, which is reflected in, among other things, thermodynamic explanations in geophysics. Boltzmann believed that macroscopic properties such as pressure and temperature can be explained by microscopic behavior of constituent particles. A Boltzmann vs. Laplacian approach to ESS would emphasize the aggregate statistical behavior of the microscopic phenomena, rather than building upwards from the most detailed scale. Whereas in Laplace's world the future is entirely determined by the present state, Boltzmann acknowledged a role for chance and probability. The Boltzmann worldview also recognizes that some uncertainty exists no matter how much information is available.

For a systems viewpoint, Austrian biologist Karl Ludwig von Bertalanffy is a good nominal representative. He is one of the founders of general systems theory (GST). GST is agnostic on causation in the Newton-Laplace-Boltzmann sense, and is a holistic framework specifically meant for the study of complex systems. It emphasizes interconnectedness, hierarchy, feedback loops, and emergence. GST also seeks "isomorphisms," or common principles and patterns across different kinds of systems. (Apologies for the all-white European male figureheads; I am happy to hear suggestions for alternatives).

If we insist on binaries, as humans are prone to do, Laplace joins Newton on the physics side, while von Bertalanffy takes rank with Darwin on the ecology side. Though Boltzmann was a physicist, his worldview as summarized here lies between the two, given the major role of entropy and thermodynamics in ecology. As I am using these scientists as "poster children" for broadly defined worldviews, the distinction does not bear detailed examination of their work. However, in this vein Gimbel (2024) presented an interesting perspective on how Boltzmann's work was influenced by Darwin.

In hydrology, a field traditionally dominated by Newtonian physics/engineering perspectives, there have been calls to develop a 'Darwinian' hydrology or to integrate Darwinian and Newtonian approaches (e.g. Harman & Troch, 2014). Among other things, these proposals focus on the value of the Darwinian emphasis on variability, history, and multiple possible evolutionary pathways.

According to Harman and Troch (2014), Darwinian hydrology focuses attention on patterns of variation (e.g., between watersheds) and seeking hypotheses to explain the variation in terms of the mechanisms and conditions that control their historical development or evolution. They suggest that some studies in paleohydrology, identification and exploration of hydrological regimes, and water balance studies exemplify Darwinian hydrology. An analogous top-down approach to the study of convection and cloud formation is outlined by Feingold and Koren (2023). The Budyko framework was labelled Darwinian as opposed to Newtonian by Sposito (2017) "because it foregoes reductionist explanations based on constitutive equations in favor of establishing universal relationships based solely on the mass and energy balance laws to which any physical system must conform". In this sense the percolation theory-based approaches outlined in Chaps. 6 and 8 could also be called Darwinian. In general, Darwinian hydrology has been interpreted as a top-down holistic approach (and thus also at least somewhat von Bertalanffian) rather that a bottom-up Laplacian/Newtonian perspective.

10.1.2 Scientific Transculturalism

Another perspective is based on scientific cultures. Enquist et al. (2024) define scientific culture as what constitutes understanding or explanation for a given problem and identify three main cultures. The *variance culture* originated in natural history and the study of qualitative variations in nature. Questions generally focus on processes that create variability. The *exactitude culture* seeks to "account for all of the known scientific understanding at an increasingly finer resolution to most faithfully and accurately capture, predict, and interrogate a system of interest" (Enquist et al., 2024). Emphasis is on including ever more detail. The *coarse-grained culture* is focused on the search for generalities that transcend variance and diversity (as in the variance culture) and that can be expressed in underlying principles or laws. Coarse-grained culture studies emphasize abstractions and simplification.

Drawbacks of the variance culture, according to Enquist et al. (2024), are that it often comes at the expense of understanding central tendencies around which variations occur, and of efforts to model, generalize, or integrate information. Exactitude culture often produces complicated, over-fitted, many-parameter models with limited transferability and disconnected to underlying principles. Disadvantages of coarse-grained approaches are possible disregard of details that could lead to poor predictions and that could forestall new lines of discovery.

Writing in the context of proposing a predictive science of the biosphere, Enquist et al. (2024) promote "scientific transculturalism" that integrates the three cultures. One example they cite is climate science. Early development of climate physics was dominated by the coarse-grained culture, applying the Navier–Stokes equations and other fundamental hydrodynamic principles to new types of oceanographic and meteorological observations developed in the variance culture. Early coarse-grained theoretical work focused on finding appropriate reductions and approximations to explain and predict weather and circulation patterns. As climate science developed, Enquist et al. explain, exactitude culture spurred development of more accurate theories, bringing in significant complications such as temperature and salinity effects on the atmosphere and ocean, and an increasingly refined understanding of vorticity and turbulence. The more complex models were tested against massive new datasets (variance culture) and were also compacted into coarse-grained perspectives that are more readily understood and checked. While the iterative structure outlined is idealized, the advance of climate science involved all three cultures and led to major advancing in weather forecasting and the understanding of paleoclimates and climate change.

Scientific transculturalism is not directly mappable to Laplace/Newton vs. Darwin (vs. Boltzmann vs. Von Bertalanffy) perspectives. Though Enquist et al. (2024) link the coarse-grained culture to Newton, Newtonian approaches (as perceived in geosciences) are prominent in exactitude culture studies.

10.1.3 Laws, Place, History

Phillips (2021) focused not on ontological or epistemological considerations of knowledge production, but on different sets of factors that affect and ultimately determine ESS. These are *laws, place,* and *history*. Laws are laws *per se*, and other principles and generalizations that are applicable, within their domains, regardless of time or place. In ESS, laws operate in specific contexts of boundary conditions, constraints, resources, and mass and energy inputs that constitute the place factors. The hydrodynamic laws governing estuary circulation, for example, must be applied in the context of tidal ranges and exchanges, hydrography, freshwater inputs, and climate and geologic frameworks. History factors reflect path dependence, the sequence of disturbances and environmental changes, and age or time for ESS development.

Real ESS (as opposed to abstract or laboratory models) are inextricably affected by laws, place, and history factors, so that the probability of existence of a given ESS state is

$$p(S_{ESS}) = \prod_{i=1}^{n} p(L_i) \prod_{j=1}^{m} p(P_m) \prod_{k=1}^{q} p(H_k). \tag{10.1}$$

The system state probability is a function of the joint probabilities of the $i = 1, 2, \ldots, n$ laws, the $j = 1, 2, \ldots, m$ local factors, and the $k = 1, 2, \ldots, q$ history factors. Even if all $p(L_i) = 1$ (this will not always be the case, due to uncertainty about which laws are applicable), the probabilities of place factors are <1 due to geographical and environmental variability, and $p(H_i) << 1$ in many cases, because histories are often unique.

The law-place-history perspective is consistent with the synthetic, integrative spirit of calls to integrate Newtonian and Darwinian approaches and scientific transculturalism. While specific studies may focus mainly or entirely on law, place, or history factors, the formalism of Eq. 10.1 emphasizes that ultimately, understanding ESS requires incorporating all three sets of factors. With respect to concerns about overfitting and transferability, Eq. 10.1 shows that the more factors included in a model or representation, the lower $p(S_{ESS})$ becomes, and thus the more specific and less general. This indicates that detailed understanding of a specific system (variance culture) would involve incorporating more factors, while maximum generality (coarse-grained culture) dictates using the fewest factors possible.

10.1.4 Dominant Controls

In the late twentieth century, hydrological models, particularly rainfall-runoff and watershed models, rapidly increased in sophistication and complexity. As suggested by Eq. 10.1, this made them less transferable or applicable to situations beyond the models' development, calibration, and testing context, or at least more difficult and expensive to transfer due to the large amount of data and information required.

Beven (2000), among others, pointed out some of these difficulties. These concerns spurred development of the *dominant processes concept* in hydrological modeling (Grayson & Blöschl, 2000; Sivakumar, 2004, 2008). Globally, given the diversity and geographical variability of watersheds, there are a nearly unlimited number of potentially relevant variables or factors. However, for any given catchment a few processes dominate hydrological response. The dominant processes concept argues that models incorporating the handful of dominant processes can be very successful, and that including additional processes or factors provides minimal improvement, often at considerable expense. The dominant processes are likely to differ between hydrological systems. The dominant processes concept thus recognizes that by accounting for key place and history factors to identify dominant processes, the appropriate relevant laws can be deployed for successful modeling. This was generalized to a more general explanatory framework beyond hydrological modeling by Phillips (2011), who called it the dominant controls concept. Other explicit applications of the dominant controls or processes concept include Seibert et al. (2017), Ran et al. (2020), McMillan et al. (2022) and Phillips (2023).

10.2 Multiple Scale Causality

If Darwin is top-down (again, a caricature, as Darwin's ideas emerged from detailed study of numerous fossils and specimens) and Laplace and Newton are bottom-up, then the multiple scale causality framework (Phillips, 2005; Inkpen, 2011; Oshan et al., 2022) could be labelled von Bertalanffian. In geosciences the phenomena of interest are affected by processes acting at both larger and smaller scales, typically ranging at least from molecular to planetary. The key controls of process-response relationships vary with spatial and temporal scale. In a hierarchy of nested scales, processes more than one hierarchical level away are not directly manifested at the scale of observation. That is, level H_i is only directly affected by levels H_{i-1} and H_{i+1}. This is the central tenet of hierarchy theory and is supported by several other theoretical and empirical arguments (Phillips, 2016, 2022).

For example, consider a problem involving modeling processes operating at fast and slow scales (subscripts *f, s*) with time steps Δt *and* spatial resolution (e.g., cell size) Δx. To ensure changes are not spatially propagated faster than they actually happen, these constraints are required:

$$\Delta t_f \leq \frac{\Delta S_f}{C} \tag{10.2}$$

$$\Delta t_s \leq \frac{\Delta S_s}{M} \tag{10.3}$$

C is the rate of the fastest and *M* of the slowest process. In numerical modelling of partial differential equation systems, these conditions correspond to the Courant-Friedrichs-Lewy (CFL) criterion, frequently used in climate models.

More generally,

$$\frac{\Delta S_s}{\Delta S_f} \leq \left(\frac{M}{C}\right)\left(\frac{\Delta t_s}{\Delta t_f}\right) \tag{10.4}$$

$$\frac{\Delta t_s}{\Delta t_f} \leq \left(\frac{C}{M}\right)\left(\frac{\Delta S_s}{\Delta S_f}\right) \tag{10.5}$$

If the spatial grid scale is fixed, for example ($\Delta S_s = \Delta S_f$), the time step for the slower processes should be longer than that of the faster processes by *C/M*. Similarly, if the time step is fixed ($\Delta t_s = \Delta t_f$), the grid size for the faster processes should be smaller than for the slower ones by a factor of *M/C*. When these criteria imply unrealistically large, small, or different grid sizes or time steps, scale independence is indicated.

From a von Bertalanffian systems perspective, ESS are often characterized by many interacting components at various scales. If the interaction matrix for the entire system is divided into sets of smaller-faster or broader-slower scale components that do not directly interact with each other, ***A, A(f), A(b)*** represent interaction matrices for the *n*-component parent system, the *m*-component system of the faster/smaller scale components, and the $q = n\text{-}m$ broader-slower scale components. λ, λ_f, λ_b, respectively, represent the eigenvalues of the parent system, the fine-scale components, and the broad-scale components. Schaffer (1981) proved that

$$\prod_{j=1}^{m} \lambda_j(f) = \left[\prod_{i=1}^{n} \lambda\right] / \left[\prod_{k}^{q} \lambda_k(b)\right] \tag{10.6}$$

$$\prod_{j=1}^{m} \lambda_k(b) = \left[\prod_{i=1}^{n} \lambda\right] / \left[\prod_{k}^{q} \lambda_j(f)\right] \tag{10.7}$$

Equations 10.6 and 10.7 show that the components operating at different scales are independent with respect to overall system behavior, indicating scale independence. Generally, if the abstracted smaller and broader scale components operate at scales $\geq$ two orders of magnitude, independence is indicated, and multiple scale causality holds.

A third perspective on multiple scale causality starts by assuming, along with Laplace, Newton, and Boltzmann, that there exists a fundamental minimum scale or level $\Delta_o > 0$. A process operating at this scale becomes observable at a broader scale $\Delta = N\Delta_o$. For instance, if the processes acting on a fluid particle are considered the fundamental physical scale, in a geoscience context they become observable at the level of flow structures as indicated by indices such as Reynolds, Rayleigh, and Peclet numbers.

Entropy of the observed pattern is directly related to that of the fundamental process:

$$H(\Delta) = N\,H(\Delta_0) \tag{10.8}$$

While adequate for simplified abstract or laboratory physical systems, Eq. (10.8) is not suitable to cases where multiple processes produce the observed phenomenon. If the observational scale is the broadest relevant scale, then entropy at that level is composed of the entropies from $i = 1, 2, \ldots, m$ smaller-scale processes observed at Δ_i.

$$H(\Delta_{\max}) = \sum_{i=1}^{m} \left[N_i H(\Delta_i) \right] \tag{10.9}$$

N_i is the scale ratio Δ/Δ_i.

Equations 10.8 and 10.9, however, assume the generating processes are independent, which is often not the case in ESS, so the equal sign is changed to $\leq$:

$$H(\Delta_{\max}) \leq \sum_{i=1}^{m} \left[N_i H(\Delta_i) \right]. \tag{10.10}$$

Now, from a top-down perspective, assume that the observational scale is the most detailed relevant level, and the observed entropy is produced by $j = 1, 2, \ldots, q$ broader scales for which independence cannot be assumed. Then

$$H(\Delta_{\min}) \leq \sum_{j=1}^{q} \left[\frac{H(\Delta_j)}{N_j} \right] \tag{10.11}$$

Combining Eqs. 10.10 and 10.11, for the typical ESS situation of causation at both larger smaller scales,

$$H(\Delta) \leq \sum_{i=1}^{m} \left[N_i H(\Delta_i) \right] + \sum_{j=1}^{q} \left[\frac{H(\Delta_j)}{N_j} \right] \tag{10.12}$$

Typically, not all processes and controls can be identified, much less measured. Let us simplify by identifying lumped or characteristic values for the m smaller scale and the q broader-scale processes, using subscripts l for local and g for global, respectively.

$$H(\Delta) \leq N_l H_l + N_g H_g \tag{10.13}$$

This framework is readily applicable to a feature mapped at multiple scales. For example, suppose the scale of interest for depth of regolith to bedrock is 1:24,000. If there exist local, detailed maps at a 1:600 scale, $N_l = 40$. If there is also a broader scale map at 1:100,000 scale, $N_g = 0.24$. The detailed maps would be sensitive to local, hillslope-scale variations, while the broader scale would reflect regional geology, topography, and other factors.

The chain rule for entropy applied to Eq. 10.13 leads to an expression for the mutual information or reduction of uncertainty in either the local or global scales by knowing the other:

$$I = \left(N_l H_l + N_g H_g \right) - H(\Delta) \tag{10.14}$$

Examples using multiscale soil mapping are presented by Phillips (2005).

The multiple scale causality perspective points to a focus on the scale or hierarchical level of interest and those *directly* relevant to it, as processes and controls at the finest and coarsest scales will be filtered through them and are thus not directly germane. Sometimes the largest or smallest scales, or phenomena near the higher or lower end of the hierarchy are the level of observation or interest, in which case the top-down or bottom-up approaches will be appropriate.

10.3 Abiotic Creativity

Successful integrated approaches to ESS are not contingent on consistency of abiotic selection and evolution with biological evolution and natural selection. However, to the extent there are commonalities, it facilitates the development of integrated approaches.

It is sometimes claimed that biota and biological evolution are creative, while physics is not (e.g., Chaitin, 2012). A distinction is also sometimes drawn between the so called 'non-adaptive complexity' of the physical environment as opposed to 'adaptive complexity' of life (Lineweaver, 2013). This raises the question of adaptation and creativity in the abiotic components of ESS. As shown in Sect. 8.4, ESS can adapt independently of their biological components, and the existence of various forms of efficiency selection in geophysical processes indicates adaptive capabilities. But can geophysical processes create novelty independently of organisms, and do successful (or at least persisting) novel features increase fitness?

Many definitions of creativity are directly connected to human endeavors, but one general definition is "the action or process of bringing something into existence or prominence," in which case creativity can potentially be associated with nonhuman biota and abiotic phenomena. Novel ESS features are defined as qualitatively distinct entities. Examples include the appearance of karst topography on carbonate rocks or redox features as Earth's atmosphere become oxygenated, or the advent of tropical cyclones as a mechanism for redistributing solar energy and heat. Novelties can also be less dramatic, such as the appearance of new landforms or hydrological flux pathways. We are interested specifically in novelties that allow or facilitate ESS adjustments or adaptations to environmental constraints, inputs, and boundary conditions.

Many examples exist of landscape elements coevolving with biota. The appearance of landforms such as stromatolites, termite mounds, and coral reefs, for instance, was driven by the evolutionary advent of calcite-secreting cyanobacteria, termites, and coral organisms. The evolutionary arrival of biomineralization genes and of bioturbation in the Cambrian and the evolution of lignin in the Paleozoic (which stimulated the deposition of coal and enabled many of the various biogeomorphic impacts of woody plants) are examples of important biological evolutionary effects on the geologic environment (Bottjer, 2017).

A comprehensive study of landform evolution in the Qiaoba-Zhongdu Danxia region of China identified several cases and episodes of geological innovation, including the phenomenon of "sedimentary breeding" as a fundamental basis for new lithologies and landforms (Wang et al., 2024). We turn now to other examples of the generation of novelties in Earth systems independent of biota.

10.3.1 Tropical cyclones and ENSO

Abiotic creativity can be considered in the VSR (variation, selection, retention or recurrence) framework. An abiotic example is the first appearance of tropical cyclones in the Pliocene. Cenozoic climate change included a long-term cooling trend leading to the development of polar ice and a steeper temperature gradient between the poles and the tropics. The Pliocene brought warmer global temperatures, producing conditions conducive to cyclogenesis and exacerbating the equator-pole heat imbalance (Federov et al., 2010). Yan et al. (2019) used model simulations to investigate the geography of tropical cyclone genesis regions in the Cenozoic. They found that conditions in the warmer Early Eocene were more favorable in the Southern Hemisphere for cyclone formation, especially the South Indian Ocean. With climate cooling, the coupled models show increasing favorability in the Northern Hemisphere for storm genesis, along with a coincident decrease in the Southern Hemisphere. The most favorable locations for cyclone formation migrated equatorward in both hemispheres. Yan et al. found a likely shift in the most favorable conditions to the western North Pacific during the Pliocene, largely due to the closure of the tropical seaways. This marks the establishment of the modern tropical cyclone distribution, which apparently persisted even during ice ages in the Pleistocene (Hobgood & Cerveny, 1988). Sedimentary and geochemical evidence of Pliocene tropical cyclones is found in the straits of Taiwan (Dashtgard et al., 2020, 2021).

Global climate conditions resulted in the emergence of tropical cyclones, which effectively redistribute heat from the tropics to the poles. By depleting the gradient, the tropical storms are favored by thermodynamic selection and have persisted.

El Niño-Southern Oscillation (ENSO) phenomena were discussed in Sect. 7.5.6. as a reflection of thermodynamic selection. The emergence of ENSO in Earth history is an innovation that can be linked to climate warming and persisted due to its role in redistributing tropical heat (Manucharyan & Federov, 2014). Pre-Pleistocene ENSO dynamics are uncertain and subject to debate (Lu et al., 2018), but evidence of ENSO appears in the geologic record in concert with past warm climate states in the Pliocene, Miocene, Eocene, and late Cretaceous (e.g., Galeotti et al., 2010; Von Der Heydt & Dijkstra, 2011; Davies et al., 2012).

10.3.2 Dolines

The evolution of dolines was shown to be an example of creation of novelty in landform evolution (Phillips, 2020). While karstification involves biota and advanced development requires a biological CO_2 subsidy from respiration, the key processes are not biologically dependent, and in any case the processes do not depend on any specific organisms or taxa. Thus, evolutionary creativity is independent of biological effects in this case. Phillips (2020) acknowledged that it was not clear whether or how the emergence of new forms enhanced adjustment, adaptation, or fitness—though this likely is the case, as karstification goes hand-in-hand with the evolution of store-and-pour systems.

Here we explore this further using a chronosequence of fluviokarst evolution presented by Phillips (2018). In the Kentucky River gorge area of central Kentucky, incised meander bends have been developing since the river began a period of incision in response to glacial reorganization of regional drainage systems about 1.5 Ma. The meander bend interiors are characterized by intense karst development. Unlike alluvial rivers, the bend interiors are not point bars but slip-off slopes with a series of strath terraces. The youngest surface is the modern active floodplain, Next is a steep valley side transition. Further up on the two bends studied in detail are three strath terraces increasing in age upward toward the upland unaffected by bend extension (Fig. 10.1). The identified landforms are shown in Table 10.1 and their occurrence on the chronosequence members in Table 10.2.

To examine the efficiency of the chronosequence we return to the cost function model of Gouy et al. (2024; see Sect. 6.5), repurposing it to examine the efficiency of the landform changes in the chronosequence, with an additional term added to reflect surface (fluvial) flow:

The cost function C is an indication of resistance and transport efficiency:

$$C = L\left(a_I C_I + a_F C_F + a_{WT} C_{WT} + a_p C_p + a_s C_s\right) \tag{10.15}$$

L is the length of a transport path between nodes (fracture intersections). The individual cost functions account for influence of the distance from inception surfaces (structural faults and horizons) C_I, effects of fractures C_F, (possible multiple) water table influences, accounting for both phreatic and vadose zones C_{WT}, karstification potential of the rock C_p, and the surface flow cost function C_s.

Following Gouy et al. (2024), (x) denotes a step function:

$$s(x) = \begin{cases} 1 - \left(1 - \dfrac{x}{d_{\max}}\right)^3 & if\; 0 < x < d_{\max} \\ 1 \; if\; x \geq d_{\max} \end{cases} \tag{10.16}$$

where d_{max} is the maximum distance over which karstification is affected by geological or hydrogeological surfaces. The cost $C_I = (d_I)$ reflects the influence of the inception surfaces with d_I the distance to the closest of those surfaces (structural faults and layers). As karstification proceeds as shown in the chronosequences, more and more of the inception surfaces are encroached upon, thus reducing the cost function C_I.

C_F represents the influence of fractures, which is zero when the features are closely aligned with fractures and > 0 otherwise. In the study area karst features are closely aligned with fractures, so C_F is assumed to be negligible and minimally influenced by development along the chronosequence.

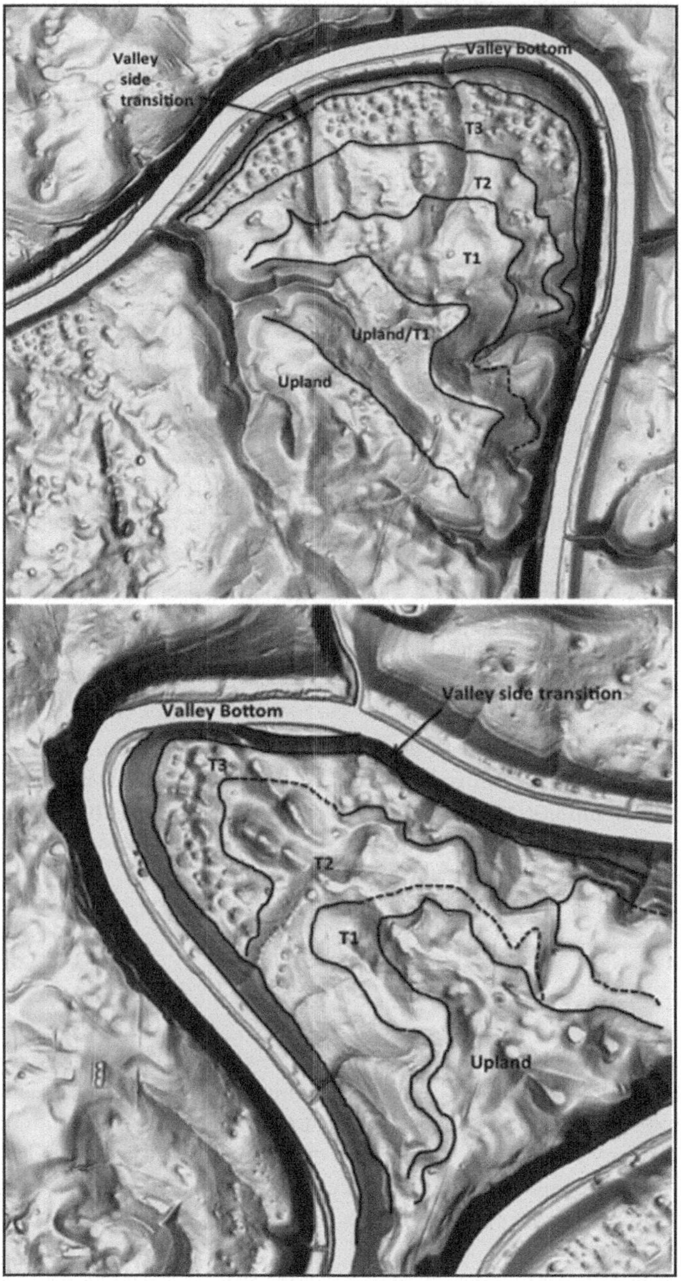

Fig. 10.1 Chronosequences at the Bowman's Bend (top) and Polly's Bend sites based on Phillips (2018). Base maps are shaded relief based on a 1.5 m resolution digital elevation model. Areas shown are 11.20 and 6.75 km^2, respectively. Coordinates at image centers are 37.8161° N, 84.6927° W (Bowman's) and 37.8022°N, 84.6472° W (Polly's)

Table 10.1 Major surface karst landforms along the study chronosequences

Code	Landform	Definition/description
ACD	Alluvial collapse doline	Sinkhole where floodplain alluvium has collapsed into dissolutional cavity in underlying limestone.
Cl	Cliff	Rock cliffs or steep bluffs at inner floodplain margin or on valley side transition.
IC	Incised channel	Perennial stream channel in V-shaped valley incised into underlying limestone, though often with alluvial cover in channel.
UIC	Unincised channel	Perennial or ephemeral stream channel in broad, low relief valley with channel confined to valley-bottom sediment.
PV	Pocket valley	Formed by collapse, or erosional exhumation of cave passages or larger conduits. Characterized by limited surface drainage areas, steep, often amphitheater-like valley heads, exposed bedrock, and groundwater seepage along joints, bedding planes, and small exposed conduits.
KV	Karst valley	Unchanneled valley formed dominantly by solution, normally dry with groundwater flow and storage beneath the valley floor.
UD	Uncovered doline	Karst sinkhole with exposed bedrock; thin, discontinuous soil cover, and steeply sloping, funnel-like base.
CD	Covered doline	Karst sinkhole mainly soil-covered except around upper rim; flat-bottomed due to sediment accumulation.
Uv	Uvala	A larger karst depression formed by the coalescence of dolines.

Table 10.2 Presence of fluviokarst landforms along chronosequences of meander bend interiors in the Kentucky River gorge

	ACD	Cl	IC	UIC	PV	KV	UD	CD	Uvala
Valley bottom (youngest)	X	X	X						
Valley side transition		X	X		X		X		
T3 (strath terrace)					X	X	X	X	
T2 (strath terrace)				X	X	X		X	X
T1 (strath terrace)				X		X		X	X
Upland (oldest)				X		X		X	X

C_{WT} incorporates water table effects, the key distinction being whether the water table is in the vadose or phreatic zone. As swallets, dolines, and other features develop and grow on the bend interiors, the depressions and general surface lowering encroach on the vadose zone and provide increasing access to conduits and vertical shafts, lowering this cost function. Over the course of landscape evolution uncovered depressions are increasingly replaced by sediment-filled ones, which increases the thickness of the vadose zone within the depressions. However, as this redistributed soil is all locally sourced within the bend, it decreases soil thickness above the epikarst in the source areas, and likely has little net effect on C_{WT}. Overall, on balance the effect of the evolution of the fluviokarst features would be expected to reduce the water table cost function.

C_p represents the karst potential of the rock. Except for older alluvial deposits at the surface, and recent ones on the floodplain, all the formations exposed in the study bends are rated as high karst potential by the Kentucky Geological Survey (https://kgs.uky.edu/kygeode/geomap/). Further, given the relatively thin alluvium overlying limestone, the low karst potential at the surface does little to inhibit collapse-type features. Therefore, the advanced fluviokarst dissection has little or no impact on this cost function.

The KarstNSim model of Gouy et al. (2024) is designed to simulate development of subsurface karst networks and thus does not include a cost function for surface flow. C_S was added to the conceptual model here, and represents the (surface) drainage density, with higher drainage densities associated with lower distance for surface runoff or epikarst flow to channels. As the development of stream valleys tends to increase, this cost function declines.

Thus, along the chronosequence all the cost functions either decrease on average or remain constant, increasing the overall efficiency. This suggests that creativity in landforms can improve evolutionary fitness via efficiency selection.

10.3.3 Sandstone Mesas

Rock masses are formed in Earth's interior under conditions of pressure and temperature orders of magnitude higher than those at the planetary surface, where conditions are relatively constant, and there is negligible exposure to H_2O and O_2. When exposed at or near the surface by uplift, volcanism, or erosional exhumation, the rocks enter a fundamentally different environment of drastically reduced pressure and temperature, exposure to water, oxygen and other corrosives, and constantly changing conditions at scales of minutes to millennia. Weathering, and subsequent transport of weathered debris, can be seen as adjustment to the surface environment.

An example of this phenomenon, explicitly framed in a landscape evolutionary context, is the work of Migon et al. (2018, 2019) on the evolution of sandstone mesas in the Elbe Mountains of eastern Germany. Using an ergodic space-for-time substitution approach (equivalent to a chronosequence), they traced the evolution of mesas from "birth" to "death."

Like all landforms, mesas have finite life spans and are reduced in area over time, transforming into buttes, and eventually into irregular arrays of boulders with the caprock completely disintegrated. "Birth" is when the mesa is separated from a plateau, and "death" when the caprock completely disintegrates. Migon et al. (2018) inferred sequential development of sandstone-capped mesas from theory, and then verified it based on field observations. The analysis indicates that mesas evolve along various pathways. As multiple pathways tend towards specific endmembers, they illustrate the principle of equifinality.

The birth-to-death sequence is summarized in Table 10.3 and Fig. 10.2.

The developmental sequence and the processes involved are clearly guided by resistance and gradient selection and produce locally novel features. However, the general forms listed in Table 10.3 do occur elsewhere. While overall the progression—which Migon et al. (2018) point out has local variations—represents increasing adjustment to the surface environment as described above, there is no available

Table 10.3 Summary of the sequential development of sandstone mesas (after Migon et al., 2019, Table 1)

Form	Description	Processes
Grosser Zschirnstein	Undissected mesa surface, with cliffs or steep slopes. Some bedding caves & blocky talus.	Bedding-controlled weathering; honeycomb weathering; block sagging; tilting and block collapse.
Gohrisch	Dissected mesa surface; continuous cliff lines; joint-aligned clefts (many roofed); detached rock towers & blocky talus.	Bedding-controlled weathering; gravity-driven cleft opening; tilting & block collapse; subsurface sand removal.
Pfaffenstein	Strongly dissected mesa surface developing into ruiniform topography; continuous cliff lines; joint-aligned clefts & canyons; residual rock towers; caves; abundant blocky talus.	Bedding-controlled weathering; cleft opening; tilting and block collapse; rock fall; rotational gliding; subsurface sand removal.
Papststein	Mesa surface reduced to narrow ridges; dry box valleys; mainly continuous cliff lines; residual rock towers; blocky talus.	Bedding-controlled weathering; cleft opening; rock fall; subsurface sand removal & sagging.
Kaisrekrone	Fragmented mesa surface; heavily weathered continuous cliffs; overhangs & joint-aligned clefts; scarce blocky talus.	Cleft opening by preferential weathering; bedding-controlled weathering
Zirkelstein	Remnant mesa surface; heavily weathered continuous cliffs; scarce blocky talus.	Honeycomb & bedding-controlled weathering; rock disintegration & sand removal.
Wolfsberg	Residual boulder pile; no cliffs or talus.	Slow gravity-driven boulder movement; subsurface sand removal.
Zeisighübel	Scattered boulders, not attached to bedrock.	Surface weathering; boulder undermining by faunalturbation.

means of assessing efficiency or cost functions as in the karst example. However, Migon et al. (2023) draw some analogies between karst and the deteriorating sandstone tablelands, citing strong surface-subsurface connectivity in ruiniform relief stages of the latter, and identify both morphological and morphogenetic analogies.

10.3.4 Genetic Algorithms

The fundamental idea of genetic algorithms is that a computer program produces variations, which are then evaluated according to some fitness criterion. If the variations improve fitness they are retained, otherwise they are discarded. Envision a metaphor of an ESS as a *K*-bit program that takes the original system *A* and produces a new variation *A'*. The fitness of the algorithm/ESS is measured by the largest integer that can be calculated, $I_{\mathrm{max.}}$ If the variation increases I_{max} it is retained and otherwise rejected. This continues up to BB(N), the "busy beaver number", which is the largest integer that can be generated by a $\leq N$-bit program that produces a single integer and then terminates. This is a direct analogy to Chaitin's (2012) algorithmic evolution model. Though for Chaitin *A* is an organism and *A'* a mutation, the

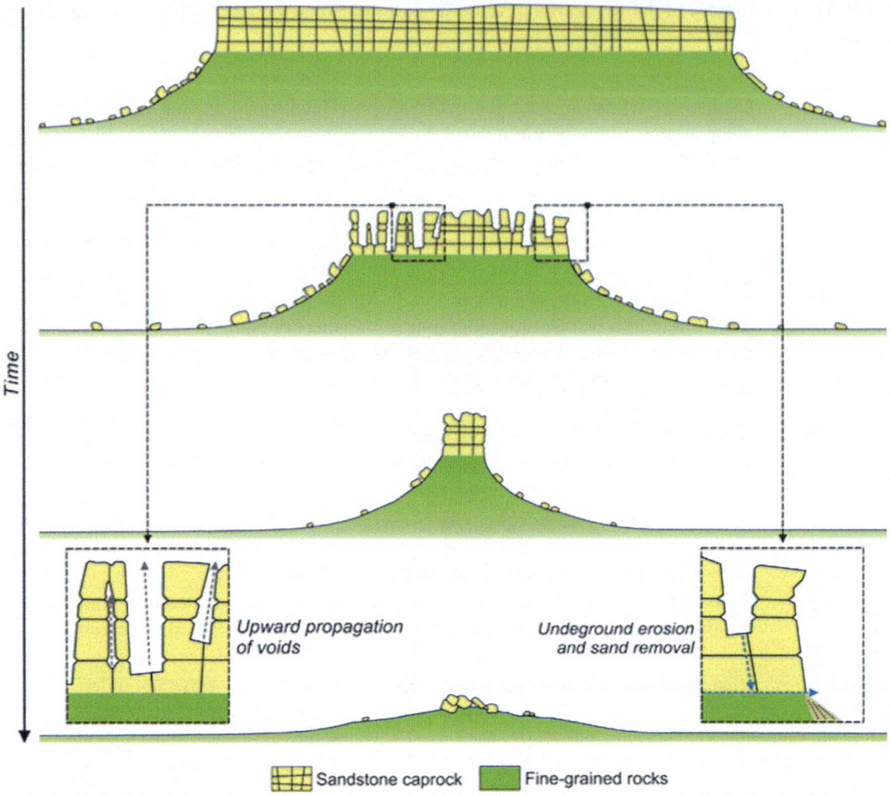

Fig. 10.2 Model of evolution of tabular sandstone hills, from plateau through mesa and butte to residual boulders. (Reprinted under CC-BY-4.0 license from Migon et al. (2019). © The Author(s))

logic is the same. Successful application of genetic algorithms (GA) indicates applicability of VSR logic to geophysical problems, and at least suggests the possibility of the emergence of fitness-improving novelties.

GA of various types have been used in many fields, including the geosciences. However, surprisingly few have been applied to the evolution of landscapes, geological entities, or Earth systems. One exception is Herman et al. (2007), who used GA to explore tectonomorphic scenarios in the Southern Alps, New Zealand. More common are studies such Feng et al.'s (2017) of the tectonic history of an area of central China. GA-based methods (An & Assumpcao, 2004) were used in the analysis of seismic waves, but selection did not play a major role in their interpretations. Similarly, a GA was applied to stable isotope records in speleothems by Pawlak and Hercman (2016), but was used to correlate records and not for their interpretation or explanation. Gouy et al. (2024) used a graph theory and GA-based approach for simulating karst networks in a way that takes account of geological controls, and briefly reviewed previous GA approaches to karst network evolution. Their analysis is discussed as an example of thermodynamic selection in Sect. 6.5.

Algorithmic models as representations of evolutionary phenomena (as opposed to computational tools) are readily utilized as an analog for biological evolution. Continuous improvement in fitness is possible in biota, up to some maximum limit presumably associated with maximum rates of biological processes. A potential argument against ever-increasing fitness or efficiency in abiotic aspects of ESS over long time periods is that these were achieved early on. For example, once climate warming in the geologic past spurred the development of the general global circulation pattern, and heat-redistributing mechanisms such as ENSO and tropical cyclones developed, there do not appear to have been any global-scale processes or features of climate produced to increase adaptation or fitness (though recent trends in cyclone climatology and other phenomena driven by global warming can be interpreted as adaptation by accelerating these mechanisms). Rather, it seems that variations in, e.g., ocean-atmosphere teleconnection patterns, ocean circulation, and the African and Asian monsoons persist as adjustments to global scale forcings, at least in the context of the post-Paleogene climate and distribution of land masses. These adjustments are guided by thermodynamic and other forms of efficiency selection, but selection has not produced any new broad-scale mechanisms since the Pliocene development of tropical cyclones.

The earliest channel networks, for another example, could develop optimal forms, though evolutionary innovations came along in concert with biological evolutionary events, such as the emergence of woody plants (Gibling et al., 2014). Then meandering and anastamosing forms developed in non-cohesive materials where they were previously absent. However, though these morphologies were new ways of increasing efficiency in non-cohesive materials, they were already developed in cohesive materials.

Organisms have DNA to pass along selected evolutionary innovations, but abiotic processes do not. Thus, a key idea in abiotic selection in ESS is *reinvention.*

10.4 Geophysical Reinvention

10.4.1 Heredity

In biological evolution, helpful and useful evolutionary innovations are passed between generations hereditarily, via genes. Organisms' DNA thus carries a memory of how to adapt to environmental conditions. Heredity is confined to lineages within a given clade or phylogenetic tree, however. Populations separated in time and space so that genetic exchange is not possible have no common heredity. Therefore, biological evolution has produced convergent evolution, where similar adaptive features have evolved in separate contexts, and divergent evolution, where different adaptations to the same opportunities and stresses arise. The results of geophysical selection, however, are nearly always convergent, with the same adaptations recurring time and again in multiple locations.

Evolution happens independently at different times and places, with the appearance of new contexts appearing continuously over geologic time owing to

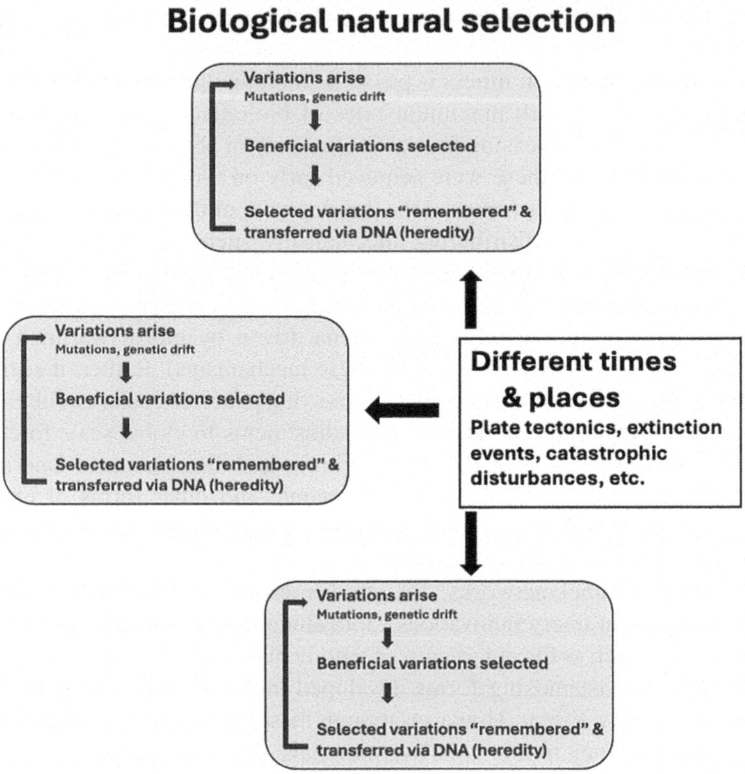

Fig. 10.3 Biological evolution in separate settings

vulcanism, tectonic movements, isostasy, extinction events, catastrophic disturbances, and sea-level changes. Figure 10.3 illustrates how the variation-selection-retention (VSR) cycle in biological evolution may occur separately over time and space. These may produce similar or different adaptations, based on recurring or new evolutionary innovations.

Geophysical and abiotic selection also occurs and recurs in new and rejuvenated settings. Within the VSR sequences, however, there is no genetic memory. Thus, selection operates from scratch, repeatedly reinventing the same general types of adjustment due to gradient, resistant, thermodynamic, and network selection (Fig. 10.4). In contrast to biological evolution, when abiotic features of ESS are destroyed or modified by disturbances, there is no hereditary memory to guide readjustment, only the general principles of selection.

10.4.2 Degrees of Freedom

In biological evolution there are more degrees of freedom for adaptation and adjustment than for abiotic selection. To give just one example, Table 10.4 outlines some

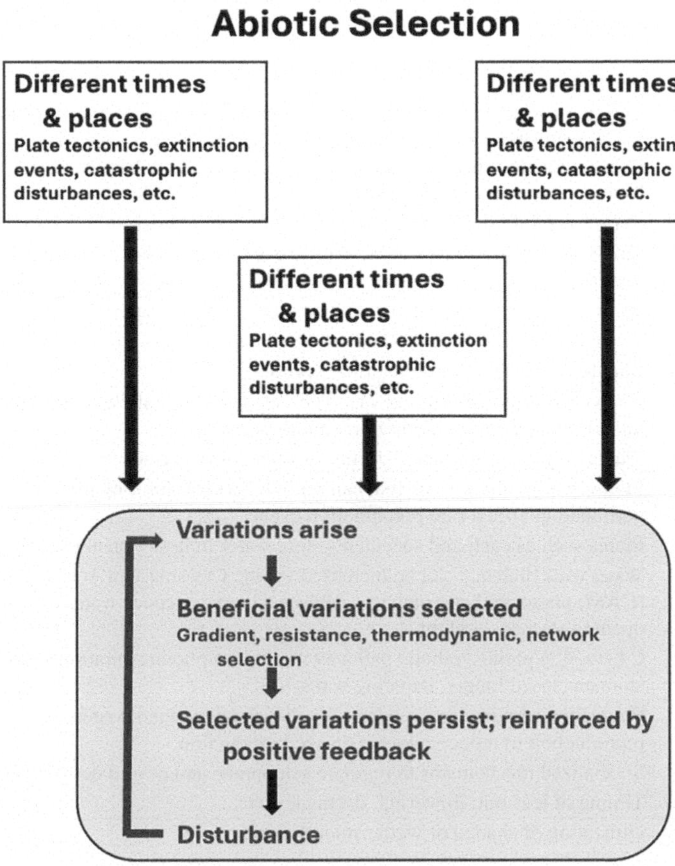

Fig. 10.4 Abiotic selection in separate settings

of the mechanisms or strategies that plants have evolved for survival in arid environments. The 17-item list is no doubt incomplete, yet gives some idea of biological evolutionary degrees of freedom.

By contrast, consider the potential adjustments of the climate system to warming (Table 10.5). Air, land, and water can heat up. Conduction and convection can transport and distribute heat in all three mediums, and state changes of water—particularly evaporation and transpiration—are driven by heat inputs. Advective movements of air masses and ocean circulation also redistribute heat energy. All of these are associated with gradient and thermodynamic selection as the system depletes gradients and equalizes energy distributions. Similarly, there are only five degrees of freedom in the ocean's response to heat accumulation (Table 10.6).

Consider also the system level of abiotic aspects of how a soil hydrologic system can respond to changes in water and energy inputs, which amounts to a half dozen mechanisms, not all of which are necessarily available in specific soil at a given time (Table 10.7). The soil water balance and soil moisture storage are the main

Table 10.4 Some potential degrees of freedom for vegetation evolutionary adaptations to aridity or water scarcity

Property	Adaptation
Size	Larger plants use more water, but often have more extensive root systems for water acquisition, more storage capacity, and ability to create shade and reduce wind to limit water loss. Smaller plants require less water and have less surface area for transpiration.
Physiognomy	Plant morphology affects surface area and microclimate.
Leaf size	Small, needle-like leaves reduce surface area and minimize transpiration losses.
Leaf texture	Waxy or leathery leaf cuticles reduce water loss by evaporation.
Leaf albedo	Lighter colored, reflective plant surfaces reduce temperature and water loss.
Leaf orientation	May be oriented to minimize sun exposure.
Root architecture	Deep roots to access groundwater (phreatophytes); shallow lateral roots to quickly soak up water during precipitation events.
Dormancy	Plants may enter dormancy during extremely dry periods to conserve water.
Seed dormancy	Plants may produce seeds that can remain dormant for long periods, germinating after a rare precipitation event.
Storage	Plants such as cacti and succulents store water in fleshy plant parts.
Photosynthesis	Water use efficiency can be increased via the Crassulacean Acid Metabolism (CAM) photosynthetic pathway, allowing them to reduce water loss by opening stomata at night.
Photosynthesis	C4 (vs. C3) photosynthetic pathway minimizes photorespiration and keeps stomata closed longer, reducing water loss.
Reproduction	Production of numerous seeds easily dispersed by wind; vegetative reproduction to reduce reliance on seed production.
Salt tolerance	Specialized mechanisms to regulate salt uptake and devoid dehydration.
Phenology	Timing of leaf-out, flowering, dormancy, etc.
Habitat specialization	Utilization of shaded or wetter microhabitats.
Life span	Ephemerals can go from germination to seed production in weeks following a rainfall event; long-lived succulents and woody species have water storage or clonal growth mechanisms that allow survival of long droughts.

Table 10.5 Climate degrees of freedom in response to increased heat

Process	Explanation
Temperature change	Heating and heat storage in water, land, and air
Heat distribution	Conduction and convection in water, land, and air
H_2O state changes	Latent heat exchanges via evapotranspiration, etc.
Advection	Air mass movements, currents, and winds to (re)distribute heat
Ocean currents	Thermohaline circulation to (re) distribute heat

Table 10.6 Ocean degrees of freedom in response to heat accumulation

Process	Explanation
Temperature change	Ocean heating
Release to atmosphere	Via evapotranspiration and convection
Ocean currents	Thermohaline circulation to (re)distribute heat
Vertical mixing	Heat distribution within water column
Atmospheric advection	Movement of ocean-derived water vapor and heat to land masses

Table 10.7 Soil hydrology degrees of freedom in response to water/energy inputs

Process	Explanation
Soil water balance	Precipitation-evapotranspiration-storage interactions
Infiltration	"Valve" controlling water input to soil vs. surface runoff
Soil moisture storage	Water available for evaporation and to plants
Percolation	Transfer of water to groundwater
Throughflow	Flow within soil matrix and/or preferential flow paths
Surface runoff and depression storage	Produced when soils saturated or precipitation rate > infiltration capacity

degrees of freedom, with the "pour" mechanisms of percolation, throughflow, and surface runoff and depression storage only potentially active during wet conditions, and infiltration functioning as a valve to partition among them.

We could, of course, break these down further, for example into the factors controlling or influencing soil moisture storage capacity. But the same is also true for the biological and other degrees of freedom.

Because abiotic aspects of ESS have fewer degrees of freedom for adjustment, even though adjustments and adaptations are reinvented *ad infinitum,* some degree of repetition in selected forms, patterns, and interrelationships is inevitable, such as dual porosity, preferential flow, dendritic channel networks, and barchan dunes. This is why abiotic selection is generally convergent.

10.5 Summary

Multiple worldviews and scientific cultures are applicable to ESS and reflected in scientific practice. These are not mutually exclusive, and no logical or practical necessity prevents the geoscientist from moving across and between them. Scientific progress, in fact, depends on us doing so, at least in the aggregate (that is, by the research community, rather than as individuals).

Some key take-home messages of this chapter:

- Newtonian, Langrangian, Darwinian, Boltzmann, and von Bertalanffian approaches have all resulted in valid, useful advances in geophysics and Earth system science. There is no reason to reject, abandon, or devalue any of these, and many good reasons to integrate them.

- The dominant controls and law-place-history frameworks are reflections of, and tools for, integration of variance, exactitude, and coarse-grained scientific cultures. Both are frameworks derived directly from Earth and environmental sciences, without reference to philosophy or the sociology, culture, or history of science.
- ESS are characterized by multiple scale causality. Thus bottom-up, top-down, and scale-specific approaches are all useful and necessary, along with explicit attention to issues of scale linkage and scale contingency. This also indicates a need to incorporate both reductionist and holistic perspectives.
- It is NOT the case that creativity resides only in biota. Abiotic processes have led to emergence of novel entities which are selected for because they enhance the efficiency or survivability of ESS. However, while biological selection may continuously produce innovations in the form of new adaptations in different time/place settings, abiotic selection repeatedly reinvents similar adaptations.
- Biological and abiotic evolution and selection have some fundamental differences, particularly with respect to the role of genes and heredity in biology. There are also some key commonalities, however, mainly based on the applicability of VSR logic. This is illustrated by the applicability of genetic and evolutionary algorithms to geophysical (as well as biological) systems.
- Climate science, writ large, and the case studies reviewed here (tropical cyclones, fluviokarst, and sandstone mesas) all illustrate abiotic creativity, and the understanding of them all relies on application of multiple, integrated worldviews.

References

An, M., & Assumpcao, M. S. (2004). Multi-objective inversion of surface waves and receiver functions by competent genetic algorithm applied to the crustal structure of the Parana Basin, SE Brazil. *Geophysical Research Letters, 31*, L05615. https://doi.org/10.1029/2003GL019179

Beven, K. J. (2000). Uniqueness of place and process representations in hydrological modeling. *Hydrology and Earth System Sciences, 4*, 203–213.

Bottjer, D. J. (2017). Geobiology and palaeogenomics. *Earth-Science Reviews, 164*, 182–192.

Chaitin, G. (2012). *Proving Darwin. Making biology mathematical* (p. 144). Pantheon.

Dashtgard, S. E., Löwemark, L., Vaucher, R., et al. (2020). Tropical cyclone deposits in the Pliocene Taiwan Strait: Processes, examples, and conceptual model. *Sedimentary Geology, 405*, 105687.

Dashtgard, S. E., Löwemark, L., Wang, P.-L., et al. (2021). Geochemical evidence of tropical cyclone controls on shallow-marine sedimentation (Pliocene, Taiwan). *Geology, 49*, 566–570.

Davies, A., Kemp, A. E. S., Weedon, G. P., & Barron, J. A. (2012). El Niño-southern oscillation variability from the late cretaceous Marca shale of California. *Geology, 40*, 15–18.

Enquist, B. J., Kempes, C. P., & West, G. B. (2024). Developing a predictive science of the biosphere requires the integration of scientific cultures. *Proceedings of the National Academy of Sciences (USA), 121*, e2209196121.

Federov, A. V., Brierley, C. M., & Emanuel, K. (2010). Tropical cyclones and permanent El Niño in the early Pliocene epoch. *Nature, 463*, 1066–1070.

Feingold, G., & Koren, I. (2023). Top-down approaches to the study of cloud systems. In Y. Liu & P. Kollias (Eds.), *Fast processes in large-scale atmospheric models: Progress, challenges, and opportunities* (Geophysical Monograph Series) (p. 13). John Wiley.

Feng, M., An, M., & Dong, S. (2017). Tectonic history of the Ordos block and Qinling Orogen inferred from crustal thickness. *Geophysical Journal International, 210*, 303–320.
Galeotti, S., von der Heydt, A., Huber, M., et al. (2010). Evidence for active El Niño southern oscillation variability in the late Miocene greenhouse climate. *Geology, 38*, 419–422.
Gibling, M. R., Davies, N. S., Falcon-Lang, H. J., et al. (2014). Palaeozoic co- evolution of rivers and vegetation: A synthesis of current knowledge. *Proceedings of the Geologists' Association, 125*, 524–533.
Gimbel, S. (2024). It ain't necessarily so: Ludwig Boltzmann's Darwinian notion of entropy. *Entropy, 26*, 238. https://doi.org/10.3390/e26030238
Gouy, A., Collon, P., Bailly-Compte, V., et al. (2024). KarstNSim: A graph-based method for 3D geologically-driven simulation of karst networks. *Journal of Hydrology, 632*, 130878.
Grayson, R. B., & Bloschl, G. (2000). Summary of pattern comparison and concluding remarks. In R. B. Grayson & G. Bloschl (Eds.), *Spatial patterns in catchment hydrology: Observations and modelling* (pp. 355–367). Cambridge University Press.
Harman, C., & Troch, P. A. (2014). What makes Darwinian hydrology "Darwinian"? *Hydrology and Earth System Sciences, 18*, 417–433.
Harte, J. (2002). Toward a synthesis of the Newtonian and Darwinian worldviews. *Physics Today, 55*(29), 2002–2034. https://doi.org/10.1063/1.1522164
Herman, F., Braun, J., & Dunlap, W. J. (2007). Tectonomorphic scenarios in the Southern Alps of New Zealand. *Journal of Geophysical Research: Solid Earth, 112*, B04201. https://doi.org/10.1029/2004JB003472
Hobgood, J. S., & Cerveny, R. S. (1988). Ice-age hurricanes and tropical storms. *Nature, 333*, 243–245.
Inkpen, R. J. (2011). Scale as a relational network in weathering studies. *Geomorphology, 130*, 10–16.
Laplace, P. S. (1814). *A philosophical essay on probabilities* (1902 English translation). Wiley.
Lineweaver, C. H. (2013). A simple treatment of complexity: Cosmological entropic boundary conditions on increasing complexity. In P. C. W. Davies, M. Ruse, & C. H. Lineweaver (Eds.), *Complexity and the arrow of time* (pp. 42–67). Cambridge University Press.
Lu, Z., Liu, Z., Zhu, J., & Cobb, K. M. (2018). A review of paleo El Niño-southern oscillation. *Atmosphere, 9*, 130.
Manucharyan, G. E., & Federov, A. V. (2014). Robust ENSO across a wide range of climates. *Journal of Climate, 27*, 5836–5850.
McMillan, H. K., Gnann, S. J., & Araki, R. (2022). Large scale evaluation of relationships between hydrologic signatures and processes. *Water Resources Research, 58*, e2021WR031751.
Migon, P., Rozycka, M., Jancewicz, K., & Duszynski, F. (2018). Evolution of sandstone mesas—Following landform decay until death. *Progress in Physical Geography, 42*, 588–606.
Migon, P., Duszynski, F., Jancewicz, K., & Rozycka, M. (2019). From plateau to plain—Using space-for-time substitution in Geoheritage interpretation, Elbsandsteingebirge, Germany. *Geoheritage, 11*, 839–853.
Migon, P., Duszynski, F., Jancewicz, K., et al. (2023). Surface-subsurface connectivity in the morphological evolution of sandstone-capped tabular hills—How much analogy to karst? *Geomorphology, 440*, 108884.
Oshan, T. M., Wolf, L. J., Sachdeva, M., et al. (2022). A scoping review of the multiplicity of scale in spatial analysis. *Journal of Geographical Systems, 24*, 293–324.
Pawlak, J., & Hercman, H. (2016). Numerical correlation of speleothem stable isotope records using a genetic algorithm. *Quaternary Geochronology, 33*, 1–12.
Phillips, J. D. (2005). Entropy analysis of multiple scale causality and qualitative causal shifts in spatial systems. *Professional Geographer, 57*, 83–93.
Phillips, J. D. (2011). Universal and local controls of avulsions in Southeast Texas rivers. *Geomorphology, 130*, 17–28.
Phillips, J. D. (2016). Vanishing point: Scale independence in geomorphic hierarchies. *Geomorphology, 266*, 66–74.

Phillips, J. D. (2018). Historical contingency in fluviokarst landscape evolution. *Geomorphology, 303*, 41–52.

Phillips, J. D. (2020). Evolutionary creativity in landscapes. *Earth Surface Processes and Landforms, 45*, 109–120.

Phillips, J. D. (2021). *Landscape evolution. Landforms, ecosystems, and soils*. Elsevier.

Phillips, J.D. (2022). The law of scale independence. *Annals of GIS 28*, 15–29. https://doi.org/10.1080/19475683.2022.2026466

Phillips, J. D. (2023). Landscape change and climate attribution, with an example from estuarine marshes. *Geomorphology, 430*, 108666.

Ran, Q., Zong, X., Ye, S., et al. (2020). Dominant mechanism for annual maximum flood and sediment events generation in the Yellow River basin. *Catena, 187*, 104376.

Schaffer, W. M. (1981). Ecological abstraction: The consequences of reduced dimensionality in ecological models. *Ecological Monographs, 51*, 383–401. https://doi.org/10.2307/2937321

Seibert, S. P., Jackisch, C., Ehret, U., et al. (2017). Unravelling abiotic and biotic controls on the seasonal water balance using data-driven dimensionless diagnostics. *Hydrology and Earth System Sciences, 21*, 2817–2841.

Sivakumar, B. (2004). Dominant processes concept in hydrology: Moving forward. *Hydrological Processes, 18*, 2349–2353.

Sivakumar, B. (2008). Dominant processes concept, model simplification and classification framework in catchment hydrology. *Stochastic Environmental Research and Risk Assessment, 22*, 737–748.

Sposito, G. (2017). Understanding the Budyko equation. *Water, 9*, 236.

Von Der Heydt, A. S., & Dijkstra, H. A. (2011). The impact of ocean gateways on ENSO variability in the Miocene. In R. Hall, R. A. Cottam, & M. E. J. Wilson (Eds.), *The SE Asian gateway: History and tectonics of the Australia-Asia collision* (Vol. 355, pp. 305–318). Geological Society of London Special Publications. https://doi.org/10.1144/SP355.15

Wang, Q., Zhang, Y., Cheng, Z., et al. (2024). Formation mechanisms of Qiaoba-Zhongdu Danxia landforms in southwestern Sichuan Province. *China. Open Geosciences, 16*, 20220709.

Yan, Q., Korty, R., & Zhang, Z. (2019). Evolution of tropical cyclone regions during the Cenozoic era. *Nature Communications, 10*, 3076.

Implications and Applications of Geophysical Selection

11

Abstract

The implications of geophysical and abiotic selection are summarized, and some applications are described in the context of global warming and climate change. Emergence guided or driven by selection explains the constrained variability of Earth systems; the simultaneous presence of unlimited variability and repeated occurrence of certain patterns, structures, and interrelationships. Tools and lessons applicable to scientific research and practical problems are reviewed. Applications to climate change focus on "climate whiplash," the increased frequency and severity of extreme precipitation and drought events, and their emerging tendency to follow closely upon one another.

11.1 Generalities and Idiosyncrasies

Think about it: all ESS are governed by some general, universal laws and principles that operate everywhere and always—either in a literal sense or within certain domains. Despite these common rules, there exists an unlimited variety of landforms, landscapes, geological formations, soils, hydrological flow systems, weather systems, ecosystems, and so on.

Now think about it another way: Earth presents a wide variety of macro- and microclimates, geological substrates, biological influences, topographic and bathymetric settings, geological histories, and so on. Yet—though no two rivers or glaciers or ocean circulation zones or rock formations are exactly the same—there exist some remarkable similarities and recurring patterns across the planet. How to explain the coexistence of these seemingly contradictory phenomena?

The constrained variation of ESS—the mixture of idiosyncrasy and similarity—is due to the operation of *selection* processes. Variations occur, and some of these are *selected* because they are more stable or efficient and others are selected against

J. D. Phillips, *Abiotic Selection in Earth Surface Systems*, Geophysics and Environmental Physics, https://doi.org/10.1007/978-3-031-85862-8_11

because they do not improve system functions or resistance. The selected-for features, patterns, pathways, and processes are retained, and often reinforced by positive feedbacks. At timescales ranging from those of weather to those of geological evolution, these phenomena favored by selection recur as they are repeatedly reinvented. This is the VSR logic of variation, selection, and retention, recurrence, or reinvention.

Gradient and resistance selection, though rarely expressed in terms of selection, have long been utilized in applications such as soil conservation and erosion and sediment control, and in assessment and management of hazards such as landslides and earthquakes. The applicability of network selection to engineering, for another example, is reflected by the fact that constructal theory arose in mechanical engineering before being applied more broadly. In some contexts, then, a perspective based on constrained selection and emergence is hardly revolutionary. However, the selection perspective facilitates placing practices in, e.g., hydraulic engineering, engineering geology, and natural resource management in a broader context of Earth surface system (ESS) evolution and functions.

In other cases, a selection perspective may open new understandings. Examples covered in this book include flow-morphology interaction in alluvial channels, where selection based on least action has enabled a consilience of theories on hydraulic geometry and channel change (Sect. 1.3.2), with obvious utility in engineering, water resource management, and flood hazards. This is also true for development of preferential flow paths and dual-porosity morphologies in soil and groundwater hydrology, and in emergence of optimal subsurface flow paths (Sects. 1.3.3, 4.1.5, 5.4, 6.3, and 8.2). Applications include agronomy, forestry, and horticulture; and water quality and pollution control. Selection-based perspectives can also produce advances in studies of natural hazards, both in terms of their processes and dynamics (e.g., slope failures, as discussed further below), and in interpreting heavy-tailed distributions of, e.g., earthquakes, floods, and extreme storms (Sect. 9.2.4).

The broad general scientific advantages of the VSR logic and selection across a broad range of fields (including the geosciences) are outlined by, e.g., Sharma and Annila (2007), Kaila and Annila (2008); Bejan (2017); Nanson and Huang (2018); Phillips (2021, 2025); and Wong et al. (2023). As discussed in Chap. 10, VSR logic and selection-based approaches are consistent with calls to integrate Darwinian and Newtonian (and Laplacian, Von Bertalanffian, and Boltzmannian) approaches.

11.2 Lessons and Tools

A common and effective guideline for engineering, planning, design, and environmental management and policy is "what would nature do?" (c.f. DeFries, 2021). What nature does is guided by selection, and affected by related phenomena such as equifinality, canalization, and adaptation. And this applies, of course, to biota, abiotic geophysical processes, and everything across and in between.

Understanding the emergent, selection-driven nature of ESS is crucial for guiding not only management and use of natural resources, but also our interpretations and reactions to environmental change. Beyond the level of individuals, biological entities cannot consciously seek any particular system state or evolutionary outcome. Abiotic elements, of course, cannot either. Thus, a view of environmental change based on supposed normal, "natural," or equilibrium standards is not realistic or helpful. When human interests seek to guide ESS development toward some outcome, or maintain a specific state, this should not be viewed as establishing a situation preferred by nature. But doing what nature would do does not always involve a hands-off approach; like many other species we modify the environment to meet human needs. But efforts to promote specific states, which are indeed often necessary or desirable, can often incorporate emergent selection according to the "what would nature do" guideline.

It is important to recognize that selection is probabilistic and local. Maximum efficiency, stability, and durability will not always be produced. And as gradient and resistance selection are local, there may be inconsistencies between local and system-scale efficiency and fitness. By contrast, network and thermodynamic selection point to the advantages of connectivity in ESS and to retention of more efficient network structures. This is relevant to assessment of impacts of network modification or destruction in ESS. Further, there is a tendency to view positive feedback in environmental systems as always deleterious, causing systems to spin out of control in unwanted ways (and this does indeed occur). But positive feedback is an important mechanism for selection in many cases, and often has beneficial effects on environmental systems (Sect. 2.2).

We have also shown that ESS can adapt in ways that do not require long time periods of incremental change via DNA transfer. This principle is intrinsically important, but can also inform human efforts to adapt to climate and other environmental changes—again answering questions of what nature would do.

As outlined earlier, both maximum entropy production (MEP) and percolation theory (PT) are closely related to selection. While tightly linked to the second law of thermodynamics, MEP is not itself a physical law; rather it is an inference method that translates physical assumptions into macroscopic predictions. Like other aspects of statistical mechanics—and like selection—MEP is probabilistic. In the ESS context, the physical assumptions imply positive selection for least action and the most rapid depletion of gradients. The successful predictions and explanations associated with MEP support abiotic selection. Conversely, the selection perspective backs the use of MEP in geoscience problems.

PT is based on underlying assumptions that are uncertain, or even patently untrue, for some ESS applications. Successful predictions from PT (and other models), by itself, does not necessarily mean that the model's underlying assumptions represent truths about nature. However, PT provides predictions based on optimal pathways for flow system efficiencies that match closely to real-world situations, and successfully predict selection independently of the method's assumptions. Abiotic selection and PT therefore also supply mutual support for each other.

A selection-based perspective therefore not only yields useful lessons about what nature does or would do, but also provides tools for identifying and implementing nature-based solutions.

In the next section we focus on implications with respect to the single most important issue facing the planet today—climate change.

11.3 Climate Change

Climate change and its numerous impacts are unfolding now, and in many cases are more severe than we had feared or predicted. Earth is perhaps entering a "new normal," but it could be more accurate to say *no* normal. That is, the environment is changing in ways that we humans have no experience with; that our planning, engineering, and economic and social policies are not prepared for and have not adapted to. Earth and environmental scientists have a special responsibility to help us adapt to the constantly shifting baseline (to the extent the baseline concept is even relevant).

Climate change is influencing, or soon will, everything we study—hydroclimatology and the hydrological cycle; storms, droughts, and other extreme weather events; natural hazards; landforms, soils, water resources, and landscape evolution. Lessons from the instrumental, observational, historical, and paleoenvironmental records will continue to be valuable, but not necessarily in drawing direct parallels between past, present, and future, as we move into uncharted territory.

What can the study of selection and interpretation of environmental observations in a selection framework offer in this regard?

We have seen in Chap. 10 that phenomena such as ENSO and tropical cyclones emerged and were selected for because they represent adaptations to global climate warming and increased temperature variations and gradients, due to thermodynamic and gradient selection. In the face of recent and ongoing climate change, there is evidence of strengthening or acceleration of these phenomena, as a selected adaptation to higher temperatures. It is therefore not unreasonable to speculate that (in addition to acceleration) new climate phenomena could emerge. This may have already happened, but we currently see them as freak occurrences or one-off responses.

For instance, "atmospheric rivers" are long narrow bands of atmospheric moisture that transport water vapor from lower to higher latitudes. They have made news in recent years by bringing heavy precipitation to the U.S. Pacific coast. Thermodynamic selection may be at the root of the intensification and poleward shift of atmospheric rivers under global warming in recent decades, as another response to increased heating and acceleration of the global hydrological cycle (Kolbe et al., 2023; Li & Ding, 2024).

Given the fundamental thermodynamic process of depleting gradients as rapidly as possible, we can suspect that any climate change impacts that affects geographical variations in energy and heat may modify, strengthen, or even create means for gradient depletion by selection of mechanisms that do this efficiently.

Some gradients, on the other hand, may be diminished. The Arctic is warming at a faster rate than the rest of the planet (Arctic amplification). The rapid warming reduces temperature gradients relative to northern hemisphere mid-latitudes. This can in turn cause the polar jet stream in the troposphere to become wavier and less stable. A more irregular jet stream can weaken the higher polar vortex in the stratosphere, a band of cold air winds that circle the Arctic. A strong vortex confines the coldest air to the Arctic, but a weakened polar vortex is subject to instability and southward shifts. These shifts, called polar or cold air outbreaks, can bring Arctic air to the midlatitudes, creating extremely cold weather, even as Earth warms overall (e.g., Zhang et al., 2016: Hamouda et al., 2024). The relationships between global climate change and the polar vortex are still being worked out. It could be useful to think of these questions in terms of whether thermodynamic selection is operating for or against the phenomena involved.

Like most Earth systems, the climate system is characterized by a dense network of feedbacks. Wunderling et al. (2024) analyzed 10 potential tipping points and nonlinear elements in the global climate system (Table 11.1), and identified 18 potential feedbacks among them, including positive and negative, stabilizing and destabilizing, and uncertain links.

Table 11.1 Potential tipping point and nonlinear elements in the global climate system. Potential tipping point and nonlinear elements in the global climate system. Data reused under CC-BY- 4.0 license from Wunderling et al. (2024). © The Author(s)

Symbol	Definition	Explanation
AMOC	Atlantic meridional overturning circulation	System of ocean currents that circulates water within the Atlantic Ocean, bringing warm water north and cold water south. It is sensitive to warming & freshening of water in the North Atlantic, the sinking of which is a key driver of the AMOC.
ENSO	El Niño-southern oscillation	See Sect. 7.5.6.
PF	Arctic permafrost	Thawing of permafrost releases meltwater in the Arctic region & releases large amounts of greenhouse gases (especially methane) to the atmosphere.
ARF	Amazon rainforest	Forest to savanna conversion & reduced tree cover impacts climate, especially forest "moisture pump."
ASI	Arctic sea ice	Warming reduces ice cover, with positive albedo effects.
WAIS	West Antarctic ice sheet	WAIS collapse would cause significant rise in sea level; freshwater release could impact AMOC.
GIS	Greenland ice sheet	GIS collapse would cause significant rise in sea level weaken AMOC via freshwater release.
ISM	Indian summer monsoon	Seasonal reversal of airflow critical to moisture regimes; could be weakened by warming.
WAM	West African monsoon	Seasonal reversal of airflow critical to moisture regimes; could be weakened by warming.
CR	Coral reefs	Severe marine ecological impacts from warming-induced coral bleaching.

Based on the feedbacks shown in Fig. 11.1 and the analysis of Wunderling et al. (2024; see also Wang, et al. 2023), I assessed the stability of the interactions among the AMOC, ENSO, WAIS, PF, ASI, and GIS (the other components in Table 11.1 do not have any outgoing feedback links and therefore do not affect the system's dynamical stability). Even by adding some self-stabilizing impacts not shown in the figure (and the well-known self-reinforcing ice-albedo feedback for ASI), and testing positive, negative, and zero feedbacks where the sign was uncertain, there was no stable configuration. Significant changes to any of the components analyzed, such as potential collapse of the AMOC due to warming waters and increased freshwater inputs to the North Atlantic, will result in a new climate state, possibly in an abrupt transition.

My stability analysis is a back-of-the-envelope calculation of a complex system with a great deal of uncertainty. But in the context of abiotic selection, it raises several key questions beyond those of decreasing the uncertainties summarized by Wang et al. (2023) and Wunderling et al. (2024). Does the dynamical instability indicated by the analysis (and suggested by the numerous potentially destabilizing individual feedbacks) represent an adjustment or adaptation mechanism of the climate system to increasing heat? Is this conceptually similar (though on far broader time and space scales) to the instability of hydrologic systems during dry periods that facilitate adaptation to changes in moisture inputs (Sects. 1.3.3. and 8.4)? How

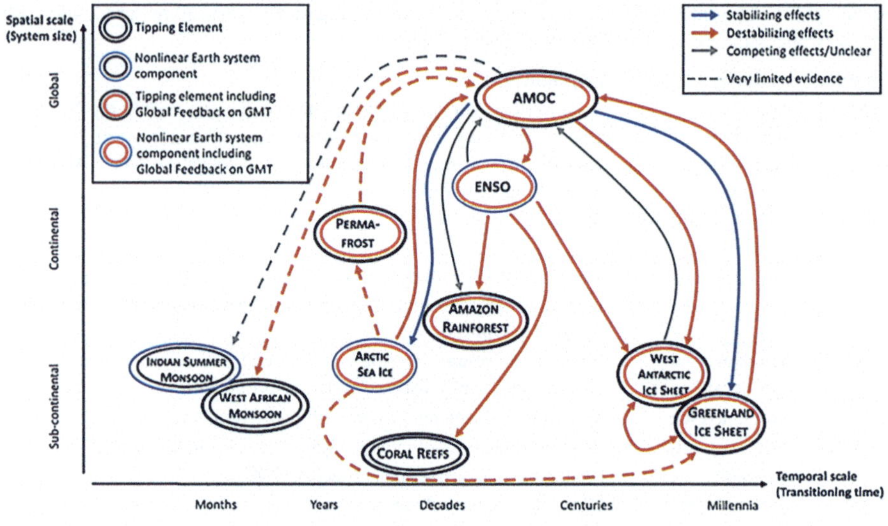

Fig. 11.1 Interactions between tipping elements across spatial and temporal scales. Temporal scales are transition times of a destabilizing tipping element.. Feedback links can be stabilizing (blue), destabilizing (red), or unclear (gray). Some tipping elements are particularly speculative (such as ENSO or the Arctic sea ice) and denoted as nonlinear Earth System components (blue border). Tipping elements that exert a feedback on the global mean temperature when they tip are depicted with an additional red inner circle. Reprinted under CC-BY- 4.0 license from Wunderling et al. (2024). © The Author(s)

do the cumulative effects fit into the thermodynamic selection driving global circulation (Sect. 7.2)? Do they support or disrupt these patterns?

We next turn to an example of an increasingly common impact of global warming, the increased frequency of extreme rainfall and droughts, often following closely upon one another.

11.3.1 Weather and Climate Whiplash

11.3.1.1 Hydrometeorological Extremes

Wetter and dryer episodes occur in almost all terrestrial environments to varying extents. Some have climatologically hard-wired wet and dry seasons, and/or susceptibility to storms. But all have at least some exposure to significantly drier or wetter periods. As recent historical and contemporary climate change continues, evidence is accumulating that the frequency and severity of both droughts and extreme rainfall events has increased and will likely continue to do so. The emerging climatology of local and regional water balances has no parallels or analogs in human or recent geological history.

Examples of unprecedented events in recent years which have been linked to trends of increasing event severity in recent decades and which bear fingerprints of human-driven climate change include the 2019–2022 drought in central-east South America, which included record soil moisture anomalies and soil desiccation (Geirenhas et al., 2023); a regional drought in the northern hemisphere winter of 2020 in Iran, Iraq, and Syria; and persistent dry, hot conditions in the summer of 2015 in India's Marathwada region (Zachariah et al., 2023).

"Flash droughts" are rapid-onset events that strain drought monitoring and forecasting capabilities. Drought intensification rates have increased in the past 60-plus years and there has been a transition to more flash droughts over about three-fourths of the globe (Yuan et al., 2023). Cook et al. (2022) reviewed exceptional dry events—megadroughts—which have occurred on every continent except Antarctica over the past ~2 Ka. Decadal-scale sea surface temperature variations are the primary driver, though radiative forcing and land-atmosphere interactions also play a role. Ongoing megadroughts in South America (Chile and Argentina) and in southwestern North America have been intensified by anthropogenic climate change. Future megadroughts are predicted to be substantially warmer than historical events, increasing megadrought severity and risk across many regions. The review highlights several gaps in paleoclimate records, incomplete process representations in climate models—and the unknown capacity of water resource managers to mitigate impacts.

Extreme high-rainfall events are also becoming more common. Summer monsoon frontal rainfall intensity in east Asia has increased in recent decades, and Moon et al. (2023) produced evidence that ongoing anthropogenic warming is a contributing factor. Their analysis found that enhanced water vapor convergence and reinforcement of the western north Pacific subtropical high increased moisture transport to the region, intensifying frontal precipitation. Another example is the severe flooding and displacement of more than 30 million people in Pakistan in the summer of 2022, while heat waves plagued central China and Europe (Hong et al., 2023).

In the southeastern U.S., Easterling et al. (2017) projected average Atlantic tropical cyclone rainfall to increase by 8–17% due to enhanced water vapor content in a warmer atmosphere. Soon after, Hurricanes Matthew, Harvey, and Florence in 2016–2018 seemed to prove them right (the analysis was conducted before the first of these storms struck in October 2016). Hurricane Harvey produced the largest single-event rainfall totals ever recorded in the U.S., with Hurricane Florence ranked seventh (Kunkel & Champion, 2019). Hurricane Matthew (2016) resulted in 24-h rainfall records at six locations in the Carolinas (Weaver et al., 2016). Hurricane Florence produced new peak streamflow records at 28 gaging stations in the region (Feaster et al., 2018). Key factors in the extreme precipitation of these storms were slow movement and exceptionally large areal coverage, both linked to recent warming (Kossin, 2019; Li & Chakraborty, 2020). As of June, 2024 many of the same areas receiving excessive rainfall in hurricanes Matthew and Florence were in severe to extreme drought, and in September 2024 Hurricane Helene, tracking inland, dumped extreme rainfall on the western Carolinas and eastern Tennessee, resulting in at least 34 new flood stage records in North Carolina alone. This is a minimum because many gaging stations failed or were destroyed while floodwaters were still rising. As this chapter was written in early November, 2024, Spain was still recovering from the largest floods on record, after enduring a prolonged drought in 2022–2023.

The North Atlantic hurricane season of 2020 was one of the most active on record, and Reed et al. (2022) linked it explicitly to climate warming. They showed that human-induced climate change caused increases in the extreme 3-h storm rainfall rates by 10% and extreme 3-day accumulated rainfall amounts by 5% for storms of at least tropical storm strength. For hurricane strength storms, extreme 3-h precipitation intensities and 3-day accumulated rainfall increased by 11 and 8%, respectively. The warming-induced increase in precipitable water was linked to increases in both total and extreme precipitation in the northeastern U.S. in recent decades (Picard et al., 2023). The increase will likely continue in the near future, they found.

11.3.1.2 Climate Whiplash

As the examples above show, some of the same regions have experienced both major droughts and extreme rainfalls in recent years. Sometimes this extremely wet to extremely dry (or vice-versa) switch can be abrupt, a phenomenon called climate (or weather) whiplash. This has been extensively discussed both in the scientific literature and media for California (e.g., Robbins, 2019; Zamora-Reyes et al., 2021; Spengelman & Osenga, 2023).

Chen and Wang (2022) addressed the consecutive occurrence of droughts and floods, including changes in frequency, intensity, and elapsed time of transitions between dry and wet periods under a warming climate. They predicted that nearly 60% of global land area will experience shorter transition times between wet and dry episodes. In southern Asia, expected to undergo the most severe dry-to-wet transitions, enhanced potential evapotranspiration (PET) variability contributes to the shortened transition times, while in southern North America warming will lead to acceleration of transition times via increases in variability of both precipitation and PET.

Rapid intraseasonal swings between heat wave plus drought and extreme wet periods are called compound whiplash events by Fang & Lu (2023). They found that global warming is likely to increase compound whiplash events by two- to 3.5-fold by 2100 under a business-as-usual climate scenario. Increasing severity and areal coverage of such events is likely, with the east Asian summer monsoon region likely to experience the most intensification. Moreover, they found, seasonality of swings with opposite directions increases due to the skewed Asian monsoon annual cycle, resulting in more frequent heat-drought to pluvial shifts in spring, and more wet to hot-dry in autumn.

Li and Rodell (2023) identified and quantified back-to-back extreme wet and dry events worldwide using a clustering technique applied to terrestrial water storage data from the Gravity Recovery and Climate Experiment (GRACE) and GRACE Follow-On. The most intense events often reflect impacts of large-scale ocean-atmosphere interactions such as ENSO, as well as consequences of climate change. They found that severity of both wet and dry events increased in most cases, probably linked to general intensification of wet and dry weather regimes in a warmer world.

In Canada, successive wet-dry and dry-wet periods analyzed using hydrologic runs analysis found that coastal areas, including the west and east coast and the Great Lakes region, are the hotspots for such abrupt transitions (RahimiMovaghar et al., 2024). Transition times between dry and wet periods were less than a month, and apparently linked to teleconnection patterns such as ENSO and the North Atlantic Oscillation. The consecutive hydro-hazards challenge resilience measures and disaster risk measures. Deng et al. (2024) evaluated the coincidence rate between extreme precipitation and drought at 1–3-month intervals for the 1985–2014 period, finding that global droughts and extreme precipitation events coincided most frequently at one-month intervals. They also project that global warming will increase the coincidence rate.

Other evidence of increasing frequency of droughts, floods, and climate whiplash, along with case study examples, is available via the World Weather Attribution project, which seeks to assess the role of climate change in extreme weather and natural hazard events (https://worldweatheratrribution.org, accessed 30 October 2024), and the Climate Attribution database (https://climateattribution.org), which as of June 2024 contained summaries of and links to nearly 700 studies, including extreme event and impact attribution research.

11.3.2 Treating Whiplash

Mimicking nature is often mentioned, with good reason, as a guideline for adapting to climate change. Another is to avoid destruction or disruption of ESS already adapted. One example was given in Sect. 6.2.4. Selection often favors highly connected networks, and it was shown how the highly connected hydrogeomorphic elements of the lower Waccamaw River allowed it to absorb the impacts of a succession of extreme floods. A recent public-private partnership effort to protect the channel wetland complexes in this region is an example of a climate adaptation strategy that also provides many other ecosystem services.

As climate change, including whiplash, is already wired in for the near future, in addition to attempts to limit further warming, the need is for *adaptation.* The example of abiotic adaptation in ESS presented in Sect. 8.4 is directly relevant here—store and pour capabilities. Resource and infrastructure management, planning, policy, engineering, hazard mitigation, and emergency response tend to be based on historical or traditional issues, or on the most recent event. That is, these programs and institutions are generally based on either drought or water scarcity or excessive moisture and floods. If capacity, plans, or policies exist for both, they are typically separate.

In nature, selection for stability encourages store and pour morphology and mechanisms in hydrological systems, with capacities for moisture storage during dry periods and for exporting (pouring) excess moisture during wet periods. Thus explains the ubiquity of dual-porosity or dual-resistance flow systems in soils, groundwater, fluvial systems, and wetlands. The general lesson for dealing with climate whiplash is to (1) include both "store," drought response and prevention and "pour" flood management and (2) integrate both within plans, policies, and agencies.

Closely related is contingent partitioning to perform ecological and geophysical jobs of hydrological systems. Selection selects for the presence of dynamical stability during excess water inputs to maintain the system as flood water is held in auxiliary storage, exported via spillway mechanisms, and detained by slow flows. Selection also favors dynamical instability during water shortages, so that contingent partitioning can react quickly to new moisture inputs.

In dryland environments store-and-pour patterns commonly take the form of hydrologically enhanced islands of vegetation cover, soil fertility, and moisture storage, nested within elements of sparsely vegetated runoff-shedding ground that exports excess water and delivers moisture to the islands. Rango et al. (2006) showed how mimicking nature by directing water toward target areas to create patterns like the natural islands can enhance functioning of ecosystems. Where such land is already severely degraded, or it is not feasible to simply let nature take its course, they showed how structures can be used to accomplish this. These options are not limited to actions requiring new inputs of capital and technology—Marquart et al. (2020) argued, based on studies in Namibia, that pastoralists can use management strategies that maintain a heterogeneous plant community, based on interactions among infiltration, moisture storage, soil fauna, and plants. Promotion or reversal of state changes can be accomplished via contingent partitioning through hydrologically mediated plant-soil interactions, including restoring grassland communities where that is beneficial (Peters et al., 2020).

Contingent partitioning and store-and-pour also come into play in environmental flow programs. In many jurisdictions, water allocations have sought to partition among domestic, industrial, and agricultural uses. Environmental or in-stream flow programs seek to expand this by considering the needs of aquatic, mainly riverine, ecosystems. Environmental flows normally address a range of discharge scenarios from flood to low flows (e.g., Praskievicz & Luo, 2020; Salinas-Rodríguez et al., 2021). Virkki et al. (2022) showed the environmental flow envelopes are increasingly transgressed worldwide and developed an approach for addressing basins where streamflow has substantially increased, and Theodoropoulus et al. (2018) found that hydrology-based environmental flow scenarios often underestimated ecological flow needs based on hydrodynamic habitat models. With increased frequency and severity of both droughts and floods, lower as well as upper bound changes will become more problematic. Contingent partitioning approaches based on capacity for adjustment and focused on hydrological and ecological (as opposed to legal and political) allocations could be highly useful in addressing these issues.

Partitioning is directly related to hydrological connectivity, and Goodwell et al. (2018) showed that ecohydrological responses to drought can be explained by changes in process connectivity, providing useful lessons for human drought response. Thus, network selection should be considered in drought and flood adaptation.

Hydrological and temperature extremes make various geological hazards more likely, such as slope failures. Landslides and other mass movements are guided and triggered by force:resistance relationships involving gradient (slope steepness) and resistance (shear strength) selection. In some cases, mass wasting may reduce the likelihood of further slope failures by (re)moving weaker slope materials and reducing slope gradients. In other situations, failures may be favorable spots for continued or future mass movements where they increase slope gradients (for example, slump headwalls), produce topographic concavities that collect water and facilitate flows and failures, or simply reposition weak material prone to failure.

For instance, Krautblatter et al. (2024) studied mega retrogressive thaw slumps, an increasingly common type of slope failure in the Arctic due to permafrost thawing caused by climate warming. These features, which involve >10^6 m^3 of mass, influence carbon budgets and biogeochemistry in the Arctic, and post serious threats to infrastructure. Using geophysical exploration methods, they found that mega retrogressive thaw slumps preferentially develop polycyclic behavior, where slumping sets the stage for future slumps. Krautblatter et al. (2024) showed that characteristics that predispose slumps to reactivation can be identified using geophysical methods.

These mega slumps are clearly (like all slope failures) governed by gradient and resistance selection, and by canalization and positive feedbacks. Additional studies using this frame of reference could help to identify potential reactivation sites where data such as electrical resistance tomography and ground penetrating radar are not available. They could also uncover principles that might help in identifying potential failure sites and associated hazards in other environments.

****

Geophysical and abiotic selection have fundamental implications for understanding Earth surface systems, and there exist numerous practical applications of the principles and perspectives. Emergence guided or driven by selection explains the constrained variability of Earth systems; the simultaneous presence of unlimited variability and repeated occurrence of certain patterns, structures, and interrelationships. These principles also provide important lessons for dealing with climate change and other environmental issues.

References

Bejan, A. (2017). Evolution in thermodynamics. Applied. *Physics Review, 4*, 011305.

Chen, H., Wang, S. (2022). Accelerated transition between dry and wet periods in a warming climate. *Geophysical Research Lettters, 49*, e2022GL099766.

Cook, B. I., Smerdon, J. E., Cook, E. R., et al. (2022). Megadroughts in the common era and the anthropocene. *Nature Review Earth and Environment, 3*, 741–757.

DeFries, R. (2021). *What would nature do? A guide for our uncertain times*. Columbia University Press.

Deng, S., Zhao, D., Chen, Z., et al. (2024). Global distribution and projected variations of compound drought-extreme precipitation events. *Earth's Future, 12*, e2024EF004809.

Easterling, D. R., Kunkel, K. E. L., et al. (2017). Precipitation change in the United States. In D. J. Wuebbles et al. (Eds.), *Climate Science Special report: Fourth National Climate Assessment. I* (pp. 207–230). U.S. Global Change Research Program.

Fang, B., & Lu, M. (2023). Asia faces a growing threat from intraseasonal compound weather whiplash. *Earth's Future, 11*, e2022EF003111.

Feaster, T.D., Weaver, J.C., Gotvald, A.J., et al., 2018. Preliminary Peak Stage and Streamflow Data at Selected Streamgaging Stations in North Carolina and South Carolina for Flooding Following Hurricane Florence, September 2018. U.S. Geological Survey Open-File Report 2018-1172.

Geirenhas, J. L., Russo, A. C., Libonati, R., et al. (2023). Combined large-scale tropical and subtropical forcing on the severe 2019-2022 drought in South America. *npj Climate and Atmospheric Science, 6*, 185.

Goodwell, A. E., Kumar, P., Fellows, A. W., Flerchinger, G. N. (2018). Dynamic process connectivity explains ecohydrologic responses to rainfall pulses and drought. *Proceedings of the National Academy of Sciences (USA), 115*, E8604-E8613, www.pnas.org/cgi/doi/10.1073/pnas.1800236115.

Hamouda, M. E., Portal, A., & Pasquero, C. (2024). Polar vortex disruptions by high latitude ocean warming. *Geophysical Research Letters, 51*, e2023GL107567.

Hong, C. C., Huang, A. Y., Hsu, H. H., et al. (2023). Causes of 2022 Pakistan flooding and its linkage with China and Europe heatwaves. *npj Climate and Atmospheric Science, 6*, 163.

Kaila, V. R. I., & Annila, A. (2008). Natural selection for least action. *Proceedings of the Royal Society A, 44*, 3055–3070. https://doi.org/10.1098/rspa.2008.0178

Kolbe, M., Sonnemans, J. P. J., Bintanja, R., et al. (2023). Impact of atmospheric rivers on future poleward moisture transport and Arctic climate in EC-Earth2. *Journal of Geophysical Research: Atmospheres, 128*, e2023JD038926.

Kossin, J. P. (2019). A global slowdown of tropical-cyclone translation speed. *Nature Climate Change, 558*, 104–107.

Krautblatter, M., Angelopoulos, M., Pollard, W. H., et al. (2024). Life cycles and polycyclicity of mega retrogressive thaw slumps in Arctic permafrost revealed by 2D/3D geophysics and long-term retreat monitoring. *Journal of Geophysical Research: Earth Surface, 129*, e2023JF007556. https://doi.org/10.1029/2023JF007556

Kunkel, K. E., & Champion, S. M. (2019). An assessment of rainfall from Hurricanes Harvey and Florence relative to other extremely wet storms in the United States. *Geophysical Research Letters, 46*, 13500–13506.

Li, L., & Chakraborty, P. (2020). Slower decay of landfalling hurricanes in a warming world. *Nature, 587*, 230–234.

Li, Z., & Ding, Q. (2024). Poleward shift of atmospheric rivers. *Science Advances, 10*, eadq0604.

Li, B., & Rodell, M. (2023). How have hydrological extremes changed over the past 20 years? *Journal of Climate, 36*, 8581–8599.

Marquart, A., Eldridge, D. J., Geissler, K., et al. (2020). Interconnected effects of shrubs, invertebrate-derived macropores and soil texture on water infiltration in a semi-arid savanna rangeland. *Land Degradation and Development, 31*, 2307–2318.

Moon, S., Utsumi, N., Jeong, J-H., et al. 2023. Anthropogenic warming induced intensification of summer monsoon frontal precipitation over East Asia. Science Advances 9, DOI:https://doi.org/10.1126/sciadv.adh4195, eadh4195.

Nanson, G. C., & Huang, H. Q. (2018). A philosophy of rivers: Equilibrium states, channel evolution, teleomatic change and the least action principle. *Geomorphology, 302*, 3–19.

Peters, D. P. C., Okin, G. S., Herrick, J. E., et al. (2020). Modifying connectivity to promote state change reversal: The importance of geomorphic context and plant–soil feedbacks. *Ecology, 101*, e03069.

Phillips, J. D. (2021). Landscape Evolution. *Landforms, Ecosystems, and Soils*. Elsevier.

Phillips, J. D. (2025). Mysterious Ways. *Contingency, Emergence, and Selection in Nature*. Oxford University Press.

Picard, C. J., Winter, J. M., Cockburn, C., et al. (2023). Twenty-first century increases in total and extreme precipitation across the Northeastern USA. *Climatic Change, 176*, 72.

Praskievicz, S., & Luo, C. H. (2020). Assessment of flow-ecology relationships for environmental flow standards: A synthesis focused on the Southeast USA. *Hydrological Sciences Journal, 65*, 571–582.

RahimiMovaghar, M., Fereshtehpour, M., & Najafi, M. R. (2024). Spatiotemporal pattern of successive hydro-hazards and the influence of low-frequency variability modes over Canada. *Journal of Hydrology, 634*, 131057.

Rango, A., Tartowski, S. L., Laliberte, A., et al. (2006). Islands of hydrologically enhanced biotic productivity in natural and managed arid ecosystems. *Journal of Arid Environments, 65*, 235–252.

Reed, K. A., Wehner, M. F., & Zarzcycki, C. M. (2022). Attribution of 2020 hurricane season extreme rainfall to human-induced climate change. *Nature Communications, 13*, 1905.

Robbins, J. (2019). Climate whiplash: Wild swings in extreme weather are on the rise. *Yale Environment360*. https://e360.yale.edu/features/climate-whiplash-wild-swings-in-extreme-weather-are-on-the-rise. Accessed 26 June 2024.

Salinas-Rodríguez, S. A., Sánchez-Navarro, R., & Barrios-Ordóñez, J. E. (2021). Frequency of occurrence of flow regime components: A hydrology-based approach for environmental flow assessments and water allocation for the environment. *Hydrological Sciences Journal, 66*, 193–213.

Sharma, V., & Annila, A. (2007). Natural process—Natural selection. *Biophysical Chemistry, 127*, 123–128.

Spengelman, S., & Osenga, E. (2023). 'Climate whiplash' is the new normal for California, experts say. *Yale Climate Connections*. https://yaleclimateconnections.org/2023/06/climate-whiplash-is-the-new-normal-for-california-experts-say/. Accessed 26 June 2024.

Theodoropolas, C., Georgalas, S., Mamassis, N., et al. (2018). Comparing environmental flow scenarios from hydrological methods, legislation guidelines, and hydrodynamic habitat models downstream of the Marathon Dam (Attica, Greece). *Ecohydrology, 11*, e2019.

Virkki, V., Alanara, E., Porkka, M., et al. (2022). Globally widespread and increasing violations of environmental flow envelopes. *Hydrology and Earth System Sciencces, 25*, 3315–3336.

Wang, S., Foster, A., Lenz, E. A., et al. (2023). Mechanisms and impacts of earth system tipping elements. *Reviews of Geophysics, 61*, e2021RG000757.

Weaver, J.C., Feaster, T.D., Robbins, J.C., 2016. Preliminary Peak Stage and Streamflow Data at Selected Streamgaging Stations in North Carolina and South Carolina for Flooding Following Hurricane Matthew, October 2016. U.S. Geological Survey Open-File Report 2016–1205.

Wong, M. L., Cleland, C. E., Arend, D., Jr., et al. (2023). On the roles of function and selection in evolving systems. *Proceedings of the National Academy of Sciences (USA), 102*, e2320223120.

Wunderling, N., von der Heydt, A. S., Aksenov, Y., et al. (2024). Climate tipping point interactions and cascades: A review. *Earth System Dynamics, 15*, 41–74.

Yuan, X., Wang, Y., Ji, P., et al. (2023). A global transition to flash droughts under climate change. *Science, 380*, 187–191.

Zachariah, M., Kumari, S., Mondal, A., et al. (2023). Attribution of the 2015 drought in Marathwada, India from a multivariate perspective. *Weather and Climate Extremes, 39*, 10056.

Zamora-Reyes, D., Black, B., & Trouet, V. (2021). Enhanced winter, spring, and summer hydroclimate variability across California from 1940 to 2019. *International Journal of Climatology, 42*, 4940–4952.

Zhang, J., Tian, W., Chipperfield, M. P., et al. (2016). Persistent shift of the Arctic polar vortex towards the Eurasian continent in recent decades. *Nature Climate Change, 6*, 1094–1099.

Index

J. D. Phillips, *Abiotic Selection in Earth Surface Systems*, Geophysics and Environmental Physics, https://doi.org/10.1007/978-3-031-85862-8

R

S

T

V

W

The manufacturer's authorised representative in the EU is Springer Nature Customer Service Centre GmbH, Europaplatz 3, 69115 Heidelberg, Germany. If you have any concerns regarding our products, please contact ProductSafety@springernature.com

Printed and bound by CPI Group (UK) Ltd, Croydon, CR0 4YY
01/10/2026
02225913-0011